ENCYCLOPAEDIA OF PLANT PROTECTION

ENCYCLOPAEDIA OF
PLANT PROTECTION

Vol. 2

B.A. Mannan

ANMOL PUBLICATIONS PVT. LTD.
NEW DELHI - 110 002 (INDIA)

ANMOL PUBLICATIONS PVT. LTD.
H.O.: 4374/4B, Ansari Road, Darya Ganj,
New Delhi-110 002 (India)
Ph.: 23278000, 23261597
B.O.: No. 1015, Ist Main Road, BSK IIIrd Stage
IIIrd Phase, IIIrd Block,
Bangalore - 560 085 (India)
Visit us at: www.anmolpublications.com

Encyclopaedia of Plant Protection

First Published, 2008
ISBN 978-81-261-3538-7 (Set)

PRINTED IN INDIA

Printed at Mehra Offset Press, Delhi.

Contents

PREFACE

Plant lovers, food producers and agricultural planners, all voice together to protect the plants from insects and diseases. Damages to the plants by insects are easily felt, but not realised fully. Every year, hundreds of people die of hunger. We can provide them with enough, food, if food crops are kept safe from the damages.

Plants, like human beings, are prone to different diseases. A diseased plant cannot yield good produces. Hence, plant protection has attained greater significance in today's world. Agricultural crops have been prone to damage, because of insects, rodents, diseases, weeds and nematodes, etc. since time immemorial.

Plant Protection continues to play a significant role in achieving targets of crop production. Major thrust areas of plant protection are promotion of integrated pest management, ensuring availability of safe and quality pesticides for sustaining crop production, despite the ravages of pests and diseases by streamlining the quarantine measures for accelerating the introduction of new high-yielding crop varieties, besides eliminating the chances of entry of exotic pests.

It is essential to have the basic knowledge about the plant pathogens, different kinds of diseases, caused by them to various crops and of course, the methods of controlling diseases to avoid enormous losses caused by the diseases. As plant protection is indispensable, job opportunities are always there for people, having basic knowledge of pests, crops and diseases.

This book is a serious endeavour in that direction. This *Encyclopaedia of Plant Protection,* in four volumes takes the stock of present state of art in plant protection. It also focuses on isolation, structure, characterisation and physiochemical and physiological properties of plants, synthesis, functions and mechanisms of action of plant protease inhibitors. It also deals with their inhibitory effects on insects and pathogens and with the resultant application of genetic engineering to plants for their protection.

This book caters to the need of knowledge hungry biologists, agriculturists, policy planners and specialists in the field of plant protection and of course the students, teachers, researchers and beyond that, all those concerned with agricultural production and its development.

Editor

1

Diseases and Ailments

- Plant Pathology or Phytopathology (phyton-plant; pathos-ailments (disease); logos-knowledge) is that branch of agricultural science which deals with the cause etiology, resulting losses and management of plant disease. The science of plant pathology has four major objectives
  1. To study the living, non-living and environmental causes of disease or disorder in plants.
  2. To study the mechanism (s) of disease development by pathogens.
  3. To study the interaction between the plant and the pathogen in relation to the overall environment.
  4. Develop systems of management of the diseases and reduce the loss of caused by plant diseases.
- Its aim is not only to destroy the enemies of plants but also to increase productivity of the crop and profits of the farmers.

Various Diseases

Plant disease are important because of the loss they cause to the grower. The loss can occur in the field or in the store and at any time between sowing and consumption of the harvest standing crop in the field is attacked by a disease and plants start drying or their capacity to yield satisfactorily is reduced and sometimes creating conditions of food shortage and famine.

The late blight of potato, a disease caused by the fungus *Phytophthora infestance) is* a famous example of what a plant disease can do to change the course of history in *1845.* This disease destroyed the potato crop of Ireland where potato constituted the staple diet of the majority in rural areas with Ireland, England and certain parts of continental Europe. Where potato was the staple food of population. There was famine in these countries especially Ireland and about 2 million people died of hunger and many more became diseased due to physical weakness. Many migrated to other land like North-America.

Wheat rust has been another disease that has appeared in epidemic form from time to time in many countries. This disease has forced the farmers in many part of the world to change their cropping pattern. Wheat has been replaced by corn (Maize) or rye because it was regularly destroyed by rust. In the year of the second world war *(1943)* Bengal had to face a serious famine due to attack of *Helminthosporium* leaf spot *(Helminthosporium oryzae).* Likewise, in the middle of nineteenth century coffee rust. *(Hemileia zeastatrix)* attack the coffee plantation in Ceylon (Sri Lanka) which was the biggest producer and exporter of coffee. During 1867-71 the coffee production was reduced to almost half and by *1893* the export was reduced to *93%.* The planters were forced to cut down coffee plants and shift to tea plantation.

These are only few examples, however, ordinarily, plant disease affect every crop, every year and in all parts of the world. This result in huge loss of potential production and money. Accurate data regarding the losses are not available from all countries because of difficulties in assessment but on the basis of number of plants affected and extraordinary decrease in yield.

In India the losses due to rusts of wheat can range from 8-10%, loose smut of wheat is estimated to cause 2-3% losses. Few years back ear cockle of wheat, a nematode disease, was reported to cause a loss of 50% in U.P. Different smut of Jowar, red rot, potato mosaic, sandal spike disease, bunchy top of banana, rice blast, blight etc. have caused huge losses in India.

Reasons of Plant Diseases: A pathogen is always, associated with a disease. The word "Pathogen" can be broadly defined as any agent or factor that incites "Pathos" or disease in an organism. Thus in strict sense the pathogens do not necessarily belong to living or animal groups. They may be non-living or in between the living and non-living. The pathogens are thus grouped under the following categories

(a) *Abiotic Factors:* These include mainly the deficiencies or excess of nutrients, light, moisture, aeration, abnormalities in soil conditions, atmospheric impurities etc. Examples of diseases caused by abiotic factors are : mango tip rot or fruit necrosis, "Khaira" disease of rice, hollow and black heart of potato.

(b) *Mesobiotic Causes:* These are disease incitants which are neither living nor non-living. They are considered to be on threshold of life.

  (1) *Viroids:* Naked infectious strands of nucleic acid-spindal tuber of potato, citrus exocortis.

  (2) *Viruses:* These are infectious agents made up of single type of nucleic acid (RNAor DNA) enclosed in a protein coat;

Instances: leaf curl of tomato and chilli etc.

(c) *Biotic Causes:* This category includes diseases caused by animate or living or cellular organism.

(1) Prokaryotes

(a) *Mollicutes:* These are wall less prokaryotes that include mycoplasma-like organism (MLOs) and spiroplasmas.

Instances: grassy shoot of sugarcane, little leaf of brinjal, sandal spike, papaya bunchy top.

(b) *Rickettsia like Bacteria (RLBs):* These are very small, sometimes sub-microscopic, walled bacteria causing such diseases as citrus greening.

(c) *True Bacteria:* Example of disease will of potato, soft rot of potato, citrus canker etc.

(2) Eukaryotes

(a) *Fungi:* Potato wart, cabbage club root, potato late blight, downy mildews and powdery mildews, rusts and smuts, red rot of sugarcane etc.

Nearly 70% of disease in any plant are incited by fungi.

(b) *Protozoa:* Hartrot of coconut palm, phloem necrosis of coffee.

(c) *Algae:* Red rust of mango, papaya.

(d) *Nematodes:* Root-knot of vegetables crops, molya disease of wheat and barley, ear cockle of wheat, citrus decline etc.

DIFFERENT AILMENTS

(1) *Non-infectious or Non-parasitic Diseases:* These are diseases (disorder) with which no animate, virus or viroid pathogen is associated. Therefore, they remain non-infectious and can not be transmitted from a diseased plant to another plant. Since no parasite is associated with these disease they are also known as non-parasitic diseases.

(2) *Infectious Diseases:* These diseases are always infectious, sometimes contagious and are transmitted from diseased to healthy plants. These diseases are caused by parasitic pathogens or viruses.

Infectious diseases are often classified according to their occurrence in following groups:

(a) *Endemic Disease:* The word "endemic" means prevalent in and confined to, a particular country, district or location and is applied to diseases of plants and animals including man. These diseases are natural to one country or part of the earth, when a disease is more or less constantly present from year to year in a moderate to severe form, in a particular country or part of earth it is called endemic to that area.

(b) *Epidemic or Epiphytotic Disease:* The term "epidemic" is derived from a Greek word meaning "among the people" and in the true sense applies those diseases of plants humans/animals which appear very virulently among a large section of the population.

To carry the same sense in the case of plant disease the term "epiphytotic" was coined. An epiphytotic disease is one which occurs widely but periodically. It may be present constantly in, the locality but assumes serious proportions only on occasions.

(c) *Sporadic Disease:* Sporadic diseases are those diseases which occur at very irregular intervals and locations and in relatively few instances.

- When a disease is prevalent throughout the country, continent or the world it is known as endemic disease.
- For practical purpose the diseases are also classified according to their symptoms.
- According to symptoms the disease may be rusts, wilts, smuts, blights, cankers, mildews, root rot, fruit rots, leaf spots etc.
- According to host plants the diseases may be grouped as cereals disease, forage crop disease, flax diseases, root crop diseases, plantation crop diseases etc.
- When a pathogen survives and spreads through soil, it is known as soil-borne disease.
- However when a disease spreads through seed. It is known as seed-borne disease.

- When the disease dispersal through the agency of air the pathogen is known as air-borne disease.

As is discussed in the introductory para, the primary aim of the plant pathologist is to control the disease and increase the productivity of the crop. In order to devise suitable control measures one has to throughly study the cause of plant disease its etiology and factors favourable for its dispersal, infection and Epidemiology. Once the life cycle of the pathogen is completely understood, the pathologist would like to hit at the weakest link of its life cycle through immunization or prophylatic means to achieve its control.

Plant Disease Control—Its Principles: Control practices are desirable only when the cost in terms of money and efforts is materially less than the loss expected from the disease so that the grower is ensured a margin of profit from plant protection measures since economically and ecologically it is not possible to completely eliminate a pathogen from the ecosystem, therefore, the word "control" is substituted with the word management. The disease management or control involves the following steps.

(a) Immunization

(Possession of qualities by the host that do not permit establishment of infection).

1. *Resistance:* Development of resistant varieties.
2. *Chemotherapy:* Development of biochemical resistance through the application of antibiotics and systemic fungicides and also through modification of host nutritions.

(b) Prophylaxis (Preventive Measures)

(1) *Protection:*

(a) *Mechanical:* By removing the diseased material or seed from the healthy lot before sowing.

(b) *Physical:* By heat treatment etc.

(c) *Cultural:* By modifying cultural operations like time of sowing depth, crop rotation, modification of environment, sanitation, roguing of diseased plants, production of disease free seed or propagated material.

(2) *Eradication:*

(a) *Crop Rotation:* Against those pathogens which persist for longer in the soil.

(b) *Sanitation:* To remove the diseased plants or residue from the field.

(c) *Alternate Host Eradication:* To eliminate the primary inoculum produced on them.

(d) *Plant Therapy:* Through fungicides heat and also surgery (in case of tree).

(e) *Chemicals:* Through fungicides and antibiotics.

(f) Biological Control: Through the activity of other micro organisms.

(3) *Legislation:* Aims at preventing entry of pathogen from infested area to non-infested areas.

Such regulations are called quarantine. The governments may have international or domestic quarantine against a particular disease.

Integration of Plant Disease Management Practices (IPDMP): The four basic requirement for management of plant diseases are: clean and healthy seeds, clean field or pathogen free soil, prevention of entry of and infection by a pathogen in a standing crop and precautions during harvesting and storage of the produce. The crop rotation is another precaution during growth of the crop periodical roguing of disease plant.

Plant Pathology Facts

1.	Father of Pathology	Anton de Berry
2.	Father of Mycology	P A. Micheli
3.	Father of Nematology	N. A. Cobb

4.	Father of Indian Pathology	E. J. Butler
5.	Bacterial technique was given by	Dr. Robert Koch
6.	Mycoplasma disease was first reported by	Do. *et.* al. and Ishiil *et. al.*
7.	First plant parasitic bacteria was reported by	T J. Burill
8.	First plant parasitic bacteria was	Fire blight of apple
9.	First plant parasitic nematode was reported by	Needham
10.	First plant parasitic nematode was	Anguina *tritici*
11.	"Stanley" did crystallization of	Viruses
12.	Late blight of potato caused by	*Phytophthora infestance*
13.	Brown spot of rice caused by	*Helminthosporium oryzae* caused Bengal femine (1845)
14.	Coffee rust occurred in	Sri Lanka 1867
15.	Maximum number of plant parasitic	Tylenchida order nematode belongs
16.	Loose smut of wheat is	Internally seed borne *(JRF 2000)*
17.	Indian phytopathological society was established by	E. J. Butler
18.	Antibiotic penicillin was discovered by	Alexander Flemming
19.	Example of obligate parasite is	*Albugo candida*
20.	Green ear disease of bajra is a	Soil and air borne diseases
21.	Life cycle of wheat rust is given by	K. C. Mehta
22.	Systemic fungicide is discovered by	Ven Schleming and Kulka
23.	Yellow vein mosaic virus was reported by	Uppal *et. al.*
24.	Tomato mosaic virus was reported by	Adolph Mayer

25.	Taphrina causes	Peach leaf curl
26.	Endoparasitic nematodes is Lichen (Algae + Fungi)	Meloidogyne
27.	Example of symbiosis is	Mycorrhiza (Fungi + Tree roots)
28.	Citrus canker	*Xanthomonas campestris var. citri*
29.	Molya disease of wheat due to	Nematode
30.	Black tip of mango is caused by	Boron deficiency
31.	Botulism is caused by species of	*Clostridium*
32.	Agar-agar is produced from	*Gelidium*

Bacterial Diseases

	Disease	*Causal Organism*
1.	Ratoon stunning disease of sugarcane (R.S.D.S)	*Cleibactor xyli*
2.	Stalk rot of maize	*Erwinia chrysanthemi pv. zeae*
3.	Tuber soft rot	*E. cartovora*
4.	Common scab of potato	*Streptomyces scabis*
5.	Citrus greening disease	*Diaphorina citri*
6.	Leaf spot of mango	*Xanthomonas campestris* pv. *Mangifera indica*
7.	Bacterial spike blight	*Clavibacter tritici*
8.	Bacterial brown rot of potato	*Pseudomonas solanacearum*
9.	Bacterial leaf streak of rice	*Xanthomonas campestris pv. oryzicola*
10.	Bacterial leaf blight	*Xanthomonas campestris pv. oryzae*
11.	Citrus canker	*Xanthomonas campestris pv. citri*
12.	Angular leaf spot	*Xanthomonas campestris pv. malvacearum*

Viral Diseases

	Disease	Causal Organism
1.	Leaf curl of papaya	*Nicotiana virus-10*
2.	Papaya Mosaic	Potex virus group
3.	Fiji disease of sugarcane	Fiji virus
4.	Mosaic of sugarcane	Sugarcane Mosaic Virus (S.M.V.)
5.	Bunchy top of banana	Banana virus-1
6.	Leaf roll of potato	Potato leaf roll virus (Luteo group of virus) *(IAS(Pre) 2002)*
7.	Mild mosaic of potato	Potato mild mosaic virus
8.	Yellow vein mosaic of okra	Yellow vein mosaic virus
9.	Rugose mosaic of potato	Potato virus x and potato virus *y*
10.	Tomato spotted wilt	Tomato spotted wilt virus (Tospo group of virus) *([AS (Pro) 2002)*
11.	Rice Tungro	Rice tungro virus (Bacilli-form virus) (Badna group) *(IAS (Pre) 2002)*

Nematode Diseases

	Disease	Causal Organism
1.	Ear cockle of wheat	*Anguina tritici* *(HAS(Pre) 2003)*
2.	Molya disease of barley	*Heterodera avenae* *(PG. Bikaner 2000)*
3.	Root knot of vegetables crops	*Meloidogyne javanica* *M. incognita* *M. arenaria*

Fungal Diseases

	Disease	Causal Organism
1.	Club root of crucifers	*Plasmodiophora brassicae*
2.	Black wart of potato	*Synchytrium endobioticum*

Contd...

	Disease	Causal Organism
3.	Damping off of seedlings	Most common generas *Phythium* *Phytophthora* *Fusarium* *Rhizoctonia* *Botrytis* *Ozonium*
4.	Brown spot disease of maize	*Physoderma zeae-maydis*
5.	Fruit rot of cucurbits	*Phythium aphanidermatum*
6.	Foot or stem rot of papaya	P. *aphanidermatum*
7.	White rust of crucifers	*Albugo candida*
8.	Green ear disease of bajra	*Sclerospora graminicola*
9.	Downy mildew of Jowar	*Sclerospora graminicola pr. andropopognis*
10.	Downy mildew of maize	*Pernosplerospora sacchari*
11.	Downy mildew of pea	*Pernospora pisi*
12.	Downy mildew of cucurbits	*Psedopernospora cubensis*
13.	Downy mildew of crucifers	*Pernospora parasitica* (PG 2000)
14.	Downy mildew *of* grape vines	*Plasmospora viticola*
15.	Soft rot of ginger	P. *myriotylum*
16.	Root rot of turmeric	*Pythium aphanidermatum*
17.	Late blight of potato	*Phytophthora infestance*
18.	Blight of Colocasia	*Phytophthora colocasiae*
19.	Seedling blight of castor	*Phytophthora parasitica*
20.	Bud rot of palm	*Phytophthora palmivora*
21.	Gumosis of citrus	*Phytophthora palmivora and P. parasitica*
22.	Blight of sesamum	*P. parasitica var. sesami*
23.	Stem galls of coriander	*Promyces macrosporus*
24.	Peach leaf curl	*Taphrina deformis*

Contd...

	Disease	*Causal Organism*
25.	Leaf spot of turmeric	*Taphrina maculans*
26.	Soft rot of apples	*Penicillium expansum*
27.	Powdery mildew of pea	*Erysiphi polygoni*
28.	Powdery mildew of cereals	*Erysiphe graminis*
29.	Powdery mildew of cucurbits	*E. cichoiacearum*
30.	Powdery mildew of grape vine	*Uncincola necator*
31.	Powdery mildew of apple	*Podasphara leucotricha*
32.	Ring spot of sugarcane	*Leptospharia sacchari*
33.	Stem rot of paddy	*Sclerotium oryzae*
34.	Ergot of cereals	*Calviceps purpuria*
35.	Ergot of bajra	*Calviceps fusiformis*
36.	False smut of rice	*Claviceps oryzae sativae*
37.	Apple scap	*Venturia inacqualis*
38.	Stem canker of fruit	*Necteria cinnabarinatrees*
39.	Stem rot of papaya	*Sclerotinia sclerotiorum*
40.	Loose smut of wheat	*Ustilago tritici U . nuda tritici (PG 2000)*
41.	Brown rust of wheat	*Puccinia recondita*
42.	Yellow rust of wheat	*P. striiformis*
43.	Black or stem rust of wheat	*P. graminis tritici*
44.	Flag smut of wheat	*U. agropyri*
45.	Leaf smut of rice	*Entyloma oryzae*
46.	Bunt of rice	*Neovossia horrida*
47.	Karnal bunt of wheat	*Tilletia tritici*
48.	Stinking smut of wheat	*Tilletia caries*
49.	Smut of sorghum	*Tolyposporium ehrenbergii*
50.	Smut of bajra	*T. penicillariae*
51.	Head smut of sorghum	*Sphacelotheca reiliana*
52.	Grain smut of sorghum	*Sphacelotheca sorghi*
53.	Smut of maize	*Ustilago maydis*

Contd...

	Disease	Causal Organism
54.	Covered smut of barley	*Ustilago hordi*
55.	Loose smut of bajra	*Ustilago nuda* (JRF1999)
56.	Smut of sugarcane	*Ustilago scitaminea*
57.	Covered smut of oats	*Ustilago kalleri*
58.	Loose smut of oats	*U. avenae*
59.	Bean rust	*Uromyces phoseoli typica*
60.	Cowpea rust	*U. phaseoli vignae*
61.	Leaf rust of coffee	*Hemelia vastatrix* and *Hemellia coffcicola*
62.	Rust of linseed	*Melampsora lini*
63.	Early blight of potato	*Alternaria solani*
64.	Alternaria leaf spot of crucifers	*Alternoria brassicae*
65.	Le7af blight of wheat	*Alternaria triticina*
66.	Leaf spot of groundnut Tikka disease	*Cercasporidium personatum or C. arachidicola* (PC 2000)
67.	Leaf spot of rice	*Helminthosporium oryzae*
68.	Blast of rice	*Pyricularia oryzae*
69.	Red rot of sugarcane	*Colletotrichum falcatum*
70.	Die back of chillies	*Colletotrichum capsici*
71.	Mango anthracnose	*Colletotrichum gloeosporioides*
72.	Anthracnose of bean	*Colletotrichum lindemuthianum*
73.	Wilt of pigeon pea	*Fusarium udum* (PC 2000)
74.	Wilt of cotton	*Fusarium oxysporium sp. vasinfectum*
75.	Wilt of linseed	*F. oxysporium (f. lini)*
76.	Panama wilt of banana	*F. oxysporium (f. cubense)*
77.	Wilt disease of sugarcane	*Cephelosporium sacchari*

Contd...

	Disease	*Causal Organism*
78.	Stalk rot	*Fusarium monilform*
79.	Blight of gram	*Ascochyta rabiei*
80.	Charcoal rot of soyabean	*Macrophomina phaseolina*
81.	Sheath blight of rice	*Rhizoctonia solani*
82.	Black scurf of potato	*Rhizoctonia solani*
Mycoplasmal Diseases		
1.	Little leaf of brinjal	Mycoplasma
2.	Purple top roll of potato	MLOs
3.	Marginal flavescence	MLOs
4.	Witches broom	MLOs *(JRF2002)*
5.	Grassy shoot of sugarcane	MLOs
6.	Sesamum phyllody	MLOs
Physiological, Non-infectious or Non-parasitic Disease		
1.	Black heart of potato	Unfavourable O_2 condition during storage
2.	Black tip or mango necrosis	Polluted gases (SO_2)
3.	Tip burn of paddy	Reduced O_2 and N_2 under submerged condition
4.	Khaira disease of rice	Zn deficiency *(JRF1999;JRF2002)*
5.	Grey speck of oat	Mn deficiency
6.	Heart rot of beets	Lack of Mn
7.	Pahala blight of sugarcane	Mn deficiency *(IAS 2000)*
8.	White bud of maize	Lack of Zn
9.	Mottle leaf of citrus	Zn deficiency *(PCS 2002)*
10.	Whiptail *in* cauliflower	Mo deficiency *(JRF2001)*
11.	Browning or hollow stem of cauliflower	Boron deficiency
12.	Top sickness of tobacco	deficiency of B
13.	Yellow spot of citrus	deficiency of Mo

Important Glossary

- Abiotic Disease: A disease due to non-parasitic cause.
- Alternate Host: One of the two kinds of plants on which a parasitic fungus must develop to complete the life cycle. A host of a pathogen growing during off season for the main host crop.
- Antibiotic: Chemical compound (Metabolite) of an organism the inhibits of other micro-organism..
- Atrophy: Under development or wasting away of a body part.
- Autoecious Fungus: A parasitic fungus that completes its entire life cycle on the same host.
- Axenic: That type of culture of an organism in which no other organisms are present.
- Antheridium: Male sex organ in fungi.
- Bactericide: A chemical that kills bacteria.
- Bactericcin: Bactericidal substance produced by certain strain of bacteria and active against one or more strains of the same or closely related species, their production is genetically controlled by extrachromosomal DNA
- Bacteriophage: A virus that infects bacteria and usually destroys them.
- Biological Control: Total or partial destruction of population of a plant pathogen by use of other organism.
- Biotechnology: The use of genetically modified organisms and or modern techniques and processes with biological systems for disease control, crop production.
- Blight: Symptoms characterized by rapid killing of leaves, flowers and stems.
- Blister: A raised lesion on the leaf surface which opens to expose spores.
- Blotch: A disease characterized by large irregular spots or blots on leaves, shoots and stems.

- Canker: A necrotic, often sunken, lesion on stem, branch, twig or fruit.
- Chlorosis: Yellowing of normally green tissue due to destruction of chlorophyll or failure of chlorophyll formation.
- Chronic Disease: A disease which persists in the host for a long period of time.
- Chronic Symptoms: Symptoms that appear over a long time.
- Contagious Disease: A parasitic disease whose pathogen can spread from plant to plant by contact.
- Circulative Viruses: Viruses that are acquired by their vectors through their mouth parts, accumulate internally, then are passed through their tissues and introduced into new plants via. the mouth part of the vector.
- Damage: Any reduction in the quantity and/or quality of yield that results from injury.
- Damping Off: Seed decay in soil or seedling or blight usually caused by soil inhabiting fungi.
- Diagnosis: Identification of the nature and cause of the disease.
- Die-back: A symptoms of disease characterized by progressive death of shoots branches and roots from the tip backward.
- Disease Assessment: Quantitative measurement of disease or the process of measuring disease quantitatively.
- Disease Intensity: General term for amount of disease present in a population.
- Disease Prevalence: Generally incidence of fields with diseased plants in a defined geographic area.
- Downy Mildew: A plant disease in which the sporangio or conidiophores and spores of the fungus appear as downy growth on the host surface.

- ELISA (Enzyme-Linked Immuno Sorbent Assay): A serological test in which one antibody carries with it an enzyme that releases a coloured compound.
- Epidemic: A widespread and severe outbreak of a disease in a large population.
- Epidemology: The study of factors affecting outbreak and spread of an infectious disease.
- Eradicant: A chemical that destroys a pathogen at its source.
- Eradication: Control of plant diseases by eliminating the pathogen after it is established or by eliminating the plant that carries the pathogen.
- Etiology: Study of science of the cause of the plant disease and nature of the causal agent.
- Exanthema: Eruption or discharge of gum or other substances from diseased tissues.
- Facultative Parasite: An organism that is usually a saprophyte but which under some conditions may become a parasite.
- Facultative Saprophyte: An organism which spends most of its life as a parasite but may become a saprophyte for sometime under certain conditions.
- Fumigant: A volatile toxic substance or gas used to disinfest certain areas or plants.
- Fumigation: The application of a fumigant for disinfestation.
- Fungicide: A compound that kills fungi.
- Gall: A swelling or overgrowth produced on a plant as a result of infection by certain pathogens.
- Gummosis: Production of gum by or in a plant tissue.
- Homothallism: The phenomenon in which a fungus produces compatable male and female gametes on the same mycelium.
- Horizontal Resistance: Partial resistance equally effective against all races of a pathogen.

- Heterothallic Fungi: Fungi producing compatable male and female gametes on physiologically distinct mycelia.
- Hyaline: Colourless, transparent.
- Hypovirulence: Reduced virulence of a pathogen stain as a result of the presence of transmissible double stranded RNA.
- Infection: The establishment of parasitic relationship of a parasite with the host.
- Injury: Damage to a plant by animal, physical or chemical agent visible or measurable symptoms and/ or signs caused by a pathogen or pest.
- Integrated Control: An approach to plant disease management that uses all available methods of control of a disease or of all or most of the diseases or pests of a crop plant for best control results but with the least cost and the least damage to the environment.
- In Vitro: In culture, outside the host.
- In Vivo: In the host.
- Intracellular: Within the cells through the cells.
- Intercellular: Between the cells.
- Leaf Spot: A self limiting lesion on the leaf.
- Local Lesion: A localized spot produced on a leaf upon mechanical inoculation with a virus.
- Latent Virus: A virus that does not induce symptom development in the host.
- Latent Infection: The state in which a host is infected but does not show symptoms.
- Mycoplasma Like Organism (MLO): The organism that other wise looks and behave like a mycoplasma but has not yet been grown in artificial culture.
- Mottle: An irregular pattern of indistinct light and dark areas.
- Mosaic: Symptoms of certain virus diseases of plants characteried by intermingled patches of normal and light green or yellowish colour.

- Molt: To cast off the cuticle; Eedysis instar.
- Mold: Any profuse or woolly fungus growth on damp or decaying matter or no surfaces of plant disease.
- Mildew: A fungal disease of plants in which the mycelium and spores of the fungus are seen as a whitish growth on the host surface. When the growth appears like a downy growth it is called downy mildew.
- When it is like a white or dull powder it is known as powdery mildew.
- Necrotic: Dead and discoloured.
- Nematicide: A chemical compound or physical agent that kills nematodes.
- Non-infectious Disease: A disease that is caused by an environmental factor, not by organisms and viruses.
- Obligate Parasite: A parasite that in nature can grow and multiply only on a living organism.
- Oomycetes: A fungus that forms oospores.
- Parasite: An organism that lives on or in another living organism and draws its nutrition from that organism.
- Pathogen: An entity that can incite disease.
- Polygenic: A character controlled by many genes.
- Proliferation: Abnormal outgrowth.
- Phytotoxic: Tox to plants.
- Pathogenicity: The ability of a pathogen to cause disease.
- Postule: A small blister-like elevation of the epidermis, often opening to expose spores.
- Quarantine: Control of import and export of plants or plant parts to prevent spread of disease and pest.
- Resistance: Ability of a host plant to supress or retard the activity of a pathogen or either injurious factors.
- Resistant: Possessing qualities that hinder the development of a given pathogen, little or no infection.
- Ring Spot: A disease symptom characterized by a circular area of chlorosis with a green centre often caused by viruses.

- Rosette: Short, bunchy habit of plant growth.
- Rot: The softening discolouration and often disintegration of a succulant plant tissue as a result of fungal or bacterial infection.
- Rust: A disease giving a rusty appearance to a plant and caused by one of the uredinales.
- Scab: A roughened crust-like disease area of the surface of a plant organ. A disease in which such areas are formed.
- Scorch: "Burning" of leaf margins as a result of infection or unfavourable environmental conditions.
- Secondary Infection: Infection caused by inoculum produced as a result of primary infection.
- Serology: A method using specificity of the antigen-antibody reaction for the detection and identification of antigenic substances and the organisms that carry them.
- Smut: A disease caused by smut fungi and characterized by appearance of shooty powdery mass on the affected organs.
- Sooty Mold: A shooty coating on foliage and fruits formed by the dark hyphae of some fungi that live on the honeydew secreted by insects.
- Sporadic: Term applied to a disease that breaks out occasionally without being constantly destructive.
- Spore: A single to many cell reproductive body of fungi and some lower plants which may develop into a new plant.
- Stem-pitting: A symptom of some virus diseases of trees characterized by depressions on the stem under the bark.
- Sterile Fungi: Fungi in which no form of spore has been seen.
- Susceptibility: The inability of a plant to resist the effect of pathogen or other injurious factors.

- Symptom: The external and internal reactions of the host to invasion of a pathogen alterations in the plant due to infection; signs of infection.
- Syndrome: A group of signs and symptoms that occur together and characterize the disease.
- Systemic: Spreading throughout the plant body internally.
- Tolerance: The ability of a plant to sustain the effects of a disease without dying or suffering serious injury or crop loss.
- Toxicity: The capability of a compound to produce injury.
- Tumour: And uncontrolled overgrowth of tissue or tissues.
- Vertical Resistance: Complete resistance to some races of the pathogen but not to other.
- Virion: A complete virus particle.
- Viroids: Small, naked, low molecular weight ribonucleic acid (RNA) that can infect plants, replicate, themselves and cause disease.
- Virus: A submicroscopic obligate parasite consisting of only nucleic acid and coat protein.
- Virulent: Capable of causing a severe disease strongly pathogenic highly aggressive.
- Wilt: Loss of rigidity and dropping of plant parts generally caused by insufficient water in plant body or by infection resulting in blockage of water transport or toxicity.
- Witches Broom: Broom like overgrowth or massed proliferation caused by dense clustering of branches of a plant.
- Yellows: A plant disease characterized by yellowing and stunting of the host plant.
- Zoospores: Spores bearing flagella and capable of movement in water.

Table

Crop	Disease or Insect	Fungicide or Insecticide Comman Name	Fungicide or Insecticide Trade Name	Remarks
Barley, Oats, Rye, Wheat	Seed decay and seeding blights including the scab fungus (seed treatments do dot control the head blight phase of scab disease)	Captan	Captan 30-DD Captan 300 Captan 400 Captan 400 DD	Thiram formulations are not registered for oats. Vitavax formulations, Raxil-Thiram, add Terra-coat LT-2 are dot registered for rye. Use according to instructions on label. Do dot use treated seed for feed or food.
		Thiram	Thiram-50WP Gustafson 42-S	
		Corboxin	Vitavax	Not for use on rye
		carboxin + ca tad	Vitavax 20-20	
		carboxin + thiram	Vitavax 200	
		difenconazole + mefanoxam	Dividend XL RTA 5-10 fl oz/100	For wheat use only. Apply pounds of seed.
		difenconazole + mefanoxam	Dividend Apron XL 0.5-1.0 fl oz/100	For wheat use only. Apply pounds of seed.
		PCNB	Gustafson LT-2N	Not for use on rye
		tebuconazole + thiram	Raxil-Thiram	

Contd...

Barley	Loose smut, Covered smut	carboxin	Vitavax Vitavax 34 Vitavax 75 W	Use according to instructions on label. Must be applied as a slurry treatment for complete coverage.
		carboxin + ca tad carboxin + thiram tebuconzole + thiram	Vitavax 20-20 Vitavax 200 Raxil-Thiram	
	Covered smut	PCNB	Terra-Coat LT-2	Use according to instructions on the label. Should be applied as a slurry treatment for complete coverage.
	Barley stripe	imazalil	Gustafson Flo-Pro IMZ	Use according to label instructions
Oats	Loose smut, covered smut	carboxin carboxin carboxin + thiram	Vitavax Vitavax 34 Vitavax 200	Use according to instructions on label. Must be applied as a slurry treatment for complete coverage.
	Loose smut	PCNB + tebuconazole + thiram	Terra-Coat Lt-2 Raxil-Thiram	Use according to instructions on the label. Should be applied as a slurry treatment for complete coverage.

Contd...

Wheat	Loose smut	Carboxin	Vitavax Vitavax 34	Use according to instructions on label. Must be applied as slurry treatment for complete coverage.
		Carboxin + ca tan	Vitavax 20-20	
		carboxin + thiram	Vitavax 200	
		difenconazole + mefanoxam	Dividend XL RTA	Apply 5-10 fl oz/100 Ib of seed.
		defenconazole + mefanoxam	Dividend Apron XL	Apply 0.5-1.0 fl oz/ 100 Ib of seed.
		tebuconazole + thiram	Raxil-Thiram	
		triadimen	Baytan 30 Flo%vable	This material is only registered for application by certified seed conditioners and is not available for farmer application. Rate is 0.75-1.5 fl oz/100 Ib of seed. Caution: Treated seed should not be planted late in the planting season or planted more 1.5 inches deep. Use according to label instructions.

Contd...

Wheat	Common bunt or stinking smut	carboxin Vitavax 34	Vitavax instruction on label. Must	Use according to be applied as a slurry treatment for complete coverage.
		carboxin + captan	Vitavax 20-20	
		carboxin + thiram	Vitavax 200	
		PCNB	Gustafson LT-2N	
		tebuconazole +	Raxil-thiram thiram	
		difenconazole + mefanoxam	Dividend XL RTA	Apply 5-10 fl oz/100 Ib of seed.
		difenconazole + mefanoxam	Dividend Apron XL	Apply 0.5-1.0 fl oz/ 100 Ib of seed.
Wheat	Common root rot complex	imazalil	Gustafson Flo-Pro IMZ	Use according to label instructions.
		difenconazole + mefanoxam	Dividend XL RTA	Apply 5-10 fl oz/100 1b of seed.
		difenconazole + mefanoxam	Dividend Apron XL	Apply 0.5-1.0 fl oz/ 100 Ib of seed.

Contd...

Wheat	Powdery	tridimenol	Baytan 30 Flowable	This fungicide will prevent the mildew overwintering of powdery mildew and leaf rust in susceptible cultivars. Rate is 1.5 fl oz/100 Ib of seed. It is only registered for available for farmer application. Caution: Treated seed should not be planted late in planting season or planted more than 1.5 inches deep. Use according to label instructions
Wheat	Leaf rust	difenconazole + mefanoxam	Dividend XL RTA	Apply 5-10 fl oz/100 Ib of seed.
		difenconazole + mefanoxam	Dividend Apron XL	Apply 0.5-1.0 fl oz/ 100 Ib of seed.
Wheat barley	Aphids (Barley Yellow Dwarf)	imidazolidinimine	Gaucho 480 Flowable	Apply 1.0 fl oz/100 Ib of seed for fall aphid control.
	Hessian fly			Apply 1.5-2.0 fl oz/100 Ib of seed for Hessian fly control.

Contd...

Leaf Rust	Apply fungicide if rust covers 1 to 3% of area of upper leaves	
Powdery Mildew	Apply fungicide if powdery mildew covers 5 to 10% of area of upper leaves	
Septoria	Apply fungicide if 25% of the indicator leaves (described below have one or more lesions.	
	Growth Stage	**Indicator Leaves**
	Jointing until flag leaf emergence:	4th and 5th leaf below the flag
	Flag leaf emergence until late boot:	flag
	Late boot until start of flowering:	3rd leaf below the flag
	Flowering until early milk stage:	2nd leaf below the flag 1st leaf below the flag

Crop	*Disease*	*Fungicide Common Name*	*Fungicide Trade Name*	*Formulated Rate/acre*	*Remarks*
Wheat and Barley	Powdery mildew on wheat and barley	Propico-nazole	Tilt 3.6 EC	4.0 fl oz	Apply only when powdery mildew covers 5-10% of area of fully expanded upper leaves. Do not apply more than 4 fl oz/a per year or after Feekes' GS 8 (flag leaf fully emerged). Do not graze or feed livestock treated forage or cut he green crop for hay or silage. After harvest, straw may be used for bedding. Follow all label instructions.
Whe at and Barley	Leaf rust on wheat and barley	Propic-o nazole	Tilt 3.6 EC	4.0 fl oz	Apply only when rust pustules cover 1-3% of area of fully expanded upper leaves. Do not apply more than 4 fl oz/a per year or after Feeke's GS 8 (flag leaf fully emerged). Do not graze or feed livestock treated forage or cut the green crop for hay or silage. after harvest straw may used for bedding. Follow all label instructions.

Contd...

Whe at	Leaf Rust	azoxyst-robin	Quardris oz	6.2-10.8 fl	Integrate into an overall disease management strotegy that includes proper selection of varieties with disease tolerance, proper timing and placement of irrigation, removal of plant debris in which inoclum overwinters, plant residue management and crop rotation. Resistance management: Do not make more than two application of Quadris per acre per. Application directions Quadris should be applied prior to or in the early stages of disease development. Application may be made at any time, immediately after jointing (Feekes 6) up to late head emergence (Feekes 10.5) A crop oil concentrate adjuvant may be added at 1.0% V/V to optimize efficacy. Do not hervest treated wheat for Forage apply later than Feekes growth stage 10.5. Do not apply more than 0.77 quarts product/acre/season (0.4 lb ai/a). Do not apply within 14 days of harvest for hay. Do not apply within 45 days of harvest for rain and straw.
Barley and rye	Barely scald, Net, blotch	Propic-onazole	Tilt 3.6 EC	4.0 fl oz	Do not graze or feed livestock treated forage or cut the green crop for hay or silage. After harvest, straw may be used for bedding. Follow all label instructions.

Contd...

Wheat	Leaf and glume blotch and Tan spot	Propico-zole	Tilt 3.6 EC	4.0 fl oz	Tilt: Do not graze or feed livestock treated forage or cut the green crop for hay or silage. After harvest, straw may be used for bedding. Follow all label instruction.
Wheat	Leaf and glume blotch and tan sopt	azoxyst robin	Quardris fl oz	6.2-10.8	Quadris should be integrated into an ovreall disease management strategy that includes proper selection of varieties with disease tolerance, proper timing and placement of irrigation, removal of plant debris in which inoculum overwinters, plant residue management and crop rotation. *Resistance Management:* Do not make more than two application of Quadris should be applied prior to or in the early stages of disease development. Application may be made at any time, immediately after jointing (Feekes 6) up to late head emergence (Feekes 10.5) A crop oil concentrate adjuvant may be added at 1% V/V to optimize efficacy. Do not apply later than Feeks Growth Stage 10.5 Do not harvest treated wheat for forage. Do not apply more than 0.77 quarts product/acre/season (0.4 lb ai/a) Do not apply within 14 days of harvest for hay. Do not apply within 45 days of harvest for grain and straw.

2

Seed Germination

Various Phases

Seed can be considered as the starting organ in the life of a higher plant. It germinates to give rise to the seedling and eventually matures into a plant. The seed is viable even if kept as such for months or even years, while other organs of plants have relatively shorter longevity. Thus, formation of seed prolongs the life of a plant.

The seed of angiosperms develops from a fertilized ovule, while the ovary itself develops into a fruit. The seed contains endosperm and cotyledons (or scutellum) and embryo. The embryo grows to form roots and shoots during germination while cotyledons or endosperm provide nourishment for the early growth of root and shoot axis. The process involves considerable amount of metabolic activities. This is preceded by the absorption of water (imbibition) by the dry seeds. A.M. Mayer and A. Poljakoff-Maber (1989) define germination as "those consecutive events which cause a dry quiescent seed in response to water uptake, to show a rise in its general metabolic activities and to initiate the formation of a seedling from the embryo".

In rather simpler terms, germination is defined as the protrusion of embryonic axis from the seed to resume plant growth. In most seeds, radicle (root) comes out first and therefore germination is frequently equated with the emergence of radicle. However, in some seeds, plumule (shoot) comes out first, for example in *Salsola.*

Chemical Composition of the Seed

Chemical composition of the seed is an important factor determining the germination of seeds and early growth of the seedlings. There is an extensive chemical interconversion during these phases of plant life.

The major components of seeds are carbohydrates, proteins and lipids. The relative abundance of these components in some seeds is given in table.

Carbohydrates: Starch is the main storage material in most important Crop seeds such as wheat, sorghum, rice and maize. Other polysaccharides such as hemicelluloses and galactomanansare also present in seeds.

Many seeds contain sugars as well. In the embryonic axis of pea seeds, sucrose accounts for 5% of the total dry weight, and in barley embryo about 14% is sucrose. Many secondary plant products are also present in seeds in the form of glycosides.

Lipids: Lipids are found in seeds both as fats and oils, depending upon the relative amounts of saturated and unsaturated fatty acids occurring as glycerides. However, most seed fatty acids are unsaturated and most commonly occurring ones are; oleic, linoleic and linolenic acids.

In addition other organic acids, both saturated and unsaturated such as acetic, butyric, palmitic, stearic, lauric and myristic acids and many others occur as glycerides. Other lipid materials found in seeds are esters up higher alcohols, sterols, phospholipids and glycolipids, tocopherols and squalene.

Chemical Composition of Seeds

Seed (species)	Percentage of dry seed		
	Carbohydrates	Proteins	Lipids
Acer saccharinum	62.0	27.5	4.0
Amygdalus communis	8.0	24.0	53.0
Arachis hypogea	20-33.0	20-30.0	40-50.0
Brassica rapa	25.0	20.0	34.0
Cannabis sativa	21.0	18.0	33.0
Chenopodium quinova	48.0	19.0	5.0
Fagopyrum esculentum	72.0	10.0	2.0
Helianthus annus	2.0	25.0	45-50.0
Linum usitatissimum	23.0	23.0	34.0
Pisum sativum	34-46.0	20.0	2.0
Ricinus communis	0	18.0	64.0
Trilkum sp.	60-75.0	13.3	2.0
Zea mays	50-70.0	10.0	5.0

Proteins: Storage proteins are generally located in protein bodies in the seeds. The composition varies according to the species. In wheat four types of proteins occur. These are: glutenins, prolamines, globulins and albumins. The glutenins and prolamins form the major components of the protein, while the globulins contribute only 6-10% of the total and the albumins 3-5%. In maize, a prolamine called "Zein" is quite abundant. In dicotyledonous plants, prolamins seem to be almost absent. Albumins and globulins are well distributed in these species, although some contain a high (upto 50% of the total protein) amount of glutenins as well. Seed storage proteins are generally rich in amino acid proline and poor in lysine, tryptophan and methionine.

Other Components: Seeds contain a large number of other substances as well. Minerals, alkaloids, free amino acids, amides, organic acids, polysterols, various phenolic compounds, growth hormones, etc. are present in seeds. In many seeds, phosphorus is stored as phytin, which is a calcium, potassium or magnesium salt of inositol hexaphosphate. In lettuce seeds, 50% of the total phosphorus occurs as phytin. Most of phytin is located in the aleurone layers of the grains.

Metabolic Changes during Germination: The metabolic rate of a dry seed is at the lowest level. However, as soon the dry seed imbibes water, the activities are revived and rejuvenated. Some important metabolic activities taking place during germination are described in the following paragraphs.

1. *Respiration:* The germination is an energy requiring process and it depends upon the resurgence of respiration. The increase in respiration is closely associated with the hydration of seeds, and it can be resolved in three successive phases in most cases. In the first phase, the respiration rate continues to increase. In the second phase, the respiration is maintained at a fairly constant level and in the third phase, there is a further sharp increase in respiration. It is believed that the hydration of seeds takes place during the first phase and the plateau is reached (second phase) when swelling has been more or less completed. The second time increase (third phase) in respiration possibly accompanies embryo (root + shoot) growth.

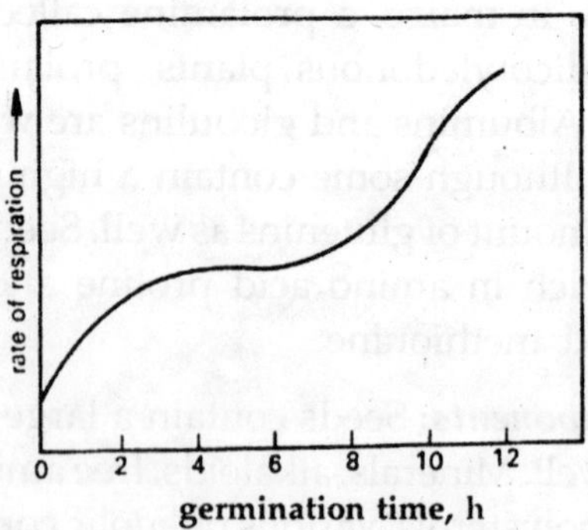

Changes in rate of respiration during seed germination.

Significant increase in ATP content and rapid conversion of AMP and ADP to ATP indicate that the oxidative phosphorylation takes place during increased respiration. In wheat embryos, ATP level and energy charge reach a plateau after 6 h of imbibition. The duration of each phase, however, differs with species and experimental conditions. External factors such as light, temperature, O_2 and CO_2 contents influence the rate of respiration.

Various inhibitors of respiration such as iodoacetate, hydroxylamine, p-nitrophenol, etc. inhibit germination. About 2-3 mM of p-nitrophenol blocks germination completely.

2. *Change in Storage Products:* During germination, the stored products such as carbohydrates, lipids and proteins are hydrolysed by the enzymes. Thus, their quantities decrease during seed germination. However, as the embryo grows, the amount of carbohydrate protein and lipids in root and shoot increases.

Starch present in the storage tissues (endosperm and cotyledons) is normally broken down by α and β-amylases. Beta amylase is present in the dry seeds, while α-amylase is synthesized during germination. The products of action of amylases are amylose, maltose and glucose. R.R. Swain and E.E Dekker (1966) demonstrate the role of another enzyme, α-glucosidase in the hydrolysis of starch in germinating pea seeds. The enzyme acts in combination with amylases as follows :

$$\text{Starch} \xrightarrow{\alpha\text{-amylase}} \text{Soluble oligosaccharides} \xrightarrow{\beta\text{-amylase}} \text{Maltose} \xrightarrow{\alpha\text{-glucosidase}} \text{Glucose}$$

Lipids are also hydrolysed to produce fatty acids and glycerol by Upases. Fatty acids are metabolised by β-oxidation and glycerol becomes part of the general carbohydrate pool present in the seed. In many seeds, such as in ground nut and castor oil seeds, the fatty acids are also converted to carbohydrate in special cell organelles called glyoxysomes.

Lipases from various species differ in their pH optima, substrate specificity and tissue and sub-cellular location.

Proteins are also hydrolysed during germination to produce amino acids and amides. The amino acids and amides are then translocated from storage tissues to the embryonic axis where they form new proteins. Germinating seeds contain a variety of proteolytic enzymes some of which are present in dry seeds while others appear during germination. The proteolytic enzymes can be divided in to two groups:

(i) Proteinases, which hydrolyse large protein molecules and

(ii) Peptidases, which hydrolyse smaller molecules usually the polypeptides.

In cereal seeds, the scutellum plays a special role in nitrogen metabolism. It acts as an intermediary between the endosperm and the growing embryo. The scutellum is capable of hydrolysing polypeptides by means of peptidases present in it.

Phosphorus is generally stored in the seeds as phytin. The phytin molecule also contains other macronutrients also such as calcium, magnesium, potassium and manganese. Orthophosphate is required for various metabolic reactions such as phospholipid synthesis and energy generating processes. The amount of phytin decreases and that of inorganic phosphate increases during germination, thus indicating that phytin acts as a reserve pool of phosphorus in the seeds. Phosphorus is liberated by the enzymic hydrolysis of phytin by a phosphatase called phytase. This enzyme is not entirely specific for phytin and is capable of hydrolysing other phosphate linkages as well.

3. *Changes in Nucleic Acids:* During germination, both cell division and cell elongation take place. It is therefore obvious that at sometime both protein synthesis and synthesis of nucleic acids also must take place.

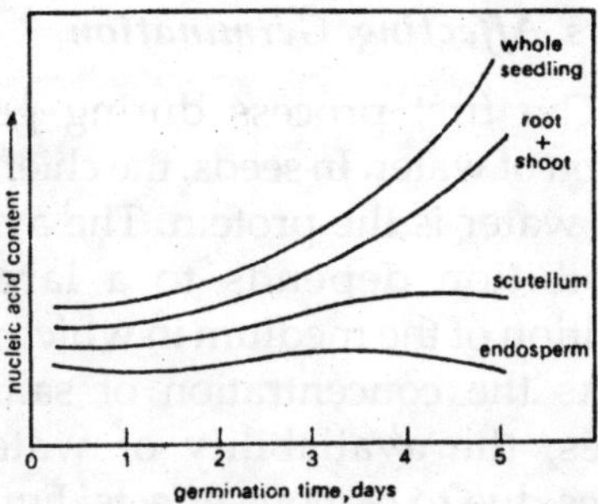

Nucleic acid content in the storage tissues hardly changes during germination. However; both DNA and RNA contents of embryonic axis increase during germination. It has been demonstrated that a long lived messenger RNA exists in many seeds, such as those of cotton, wheat, radish, *Vigna*, etc. The precise nature and function of this m-RNA is in dispute. In addition, dry seeds of rye, wheat and peas contain cytoplasmic ribonucleoproteins also. In a typical study, as done by John Ingle and others (1964) in maize seeds, total nucleic acid content of endosperm and scutellum remains almost unchanged for upto 120 h of germination. However, the nucleic acid content of the root+shoot and consequently of the entire seedling increases considerably after 48 h.

As is the case with nucleic acids, soluble nucleotide content of the embryonic axis and of whole seedling increases after 48 h, while that of scutellum and endosperm remains unchanged.

Factors Affecting Germination

Most important factor determining the germination of seeds is its own viability and life span, which of course is determined by its genetic make up. The environmental conditions under which a seed is stored also influences its viability. The seeds vary in their viability period. The most extreme case of retention of viability is that of lotus *(Nelumbo nucifera)* seeds. They are estimated to have a life of over 400 years.

Most seeds lose their viability during storage because of biochemical, ultrastructural and morphological changes.

External Factors Affecting Germination

1. *Water:* The first process during germination is the imbibition of water. In seeds, the chief component which imbibes water is the protein. The availability of water for imbibition depends to a large extent on the composition of the medium in which germination takes place. As the concentration of salts in the medium increases, the availability of water for imbibition decreases due to osmotic effects. Further, the presence of seed coat is a major barrier in the absorption of water from the medium. The seeds which have an impermeable seed coat will not swell, even if other conditions for imbibition are favourable.

Changes in Percentage Germination of Some Seeds During Storage

Seed	Years of storage		
	More than 70% germination	*30-60% germination*	*Less than 30% germination*
Avena saliva	11	12	—
Daucus carola	1	7	15
Helianlhus annus	—	—	9
Lactuca saliva	3	—	5
Medicago saliva	—	11	—
Nicoliana labaccum	—	—	11
Pisum sativum	3	—	—
Secale cereale	7	—	12
Solanum melongena	5	7	10
Spinacea oleracea	—	—	5
Triticum (wheat)	9	—	13

In field conditions, other factors such as binding of water by soil colloids, capillary forces and soil composition and texture may also influence the availability of water for germination.

2. *Atmospheric Composition:* Germination is greatly influenced by the composition of the atmosphere. Oxygen is necessary for germination, as aerobic respiration is an important physiological contributor of the process. Most seeds germinate well in air containing 20% O_2. The germination of many seeds such as those of *Xanthium* and certain cereals increases with the increase in O_2 above 20%, while those of *Typha latifolia* and *Cynodon dactylon* germinate better in the presence of 8% O_2. Some seeds such as rice and *Ethinocloa* can germinate and grow under anaerobic conditions.

The effect of CO_2 is usually the reverse of O_2. Most seeds fail to germinate if the CO_2 content is increased. However, there are other variations in the effect of CO_2 on germination of seeds, such as:

(1) There is no effect of high CO_2 content, if O_2 content is unchanged. Example; *Atriplex*.

(2) There is increase in germination with increasing CO_2 content. Example; *Phleum pratense*.

Presence of ethylene and many other pollutant gases in the atmosphere also influences germination. Low concentration of ethylene generally promotes germination of some seeds. Endogenous ethylene formation during germination also seems to be having some function in the germination. Ethylene interacts with growth hormones and its effect is regarded as that of a growth regulator.

Many other atmospheric pollutants such as NO_2, SO_2, O_3, NH_3, H_2S and p are also known to influence germination. Most of these inhibit germination, although some pollutants (such as NO_2) increase germination at very low concentrations, in some cases.

3. *Temperature:* The optimum temperature for seed germination varies according to the species, varieties and seed conditions. In most cases, seeds fail to germinate at lower temperatures. Germination is also checked above a certain temperature range. For most

tropical species, 25-35°C seems to be optimum temperature.

The effect of temperature on seed germination seems to be related to the changes in membrane structure and function and in protein denaturation. S.B. Hendricks and R.B. Taylorson (1979) have demonstrated that the properties of seed membranes changed with changes in temperature and that leakiness of seeds increased at high temperatures.

Temperature seems to be interacting with light and humidity also as far as its effect on germination is concerned. For example, seeds of *Veronica longifolia* and *Epibolium hirsutum* germinate better at lower temperature in light than in the dark. For good germination in dark a higher temperature treatment is required.

4. *Light:* The seeds of most cultivated plants germinate equally well in the dark and in the light. However, the seeds of wild plants show variability in response to light/dark. There are three categories of seeds as far as the response of germination to light is concerned. These are:

   (1) Seeds whose germination is promoted by light (positive photoblastic). Examples are—*Alisma plantago, Capparis spinosa, Daucus carota, Fagus sylvatica, Lactuca sativa, Lythrum ringens, Nicotiana tabaccum, Rumex crispus, Veronica arvensis,* etc.

   (2) Seeds whose germination is promoted by dark (retarded by light and hence negative photoblastic). Examples are—*Datura stramonium, Forsythia suspensa, Gladiolus communis, Mirabilis jalapa, Nigella* sp., *Phacelia* sp., etc.

   (3) Seeds whose germination is indifferent to light or dark. Examples are—*Anemone nemorosa, Bryonia alba, Sorghum* sp., etc.

The quality of light also influences the rate of germination. In lettuce seeds, the germination is stimulated by red light and inhibited if the seeds are subsequently illuminated with far-red light. Further illumination with red light induces germination. From these observations, it has been concluded that the effect of light on seed germination is phytochrome mediated. The presence of phytochrome and its interconversion has been adequately documented now.

The inhibition of seed germination by white light in many cases is also due to far-red irradiation. It appears that an intermediate (Between Pr and Pfr) species of phytochrome is also formed in these seeds. This phytochrome is quite reactive and may be the cause of inhibition.

The exact mechanism of induction of germination by active form of phytochrome is not very clear. Phytochrome is believed to be located in or near membranes. According to a model proposed by A.M. Mayer (1986), the induction involves the transport of Ca^{2+} and the involvement of the calcium binding protein, the calmodulin. The calmodulin mediates primary/secondary responses, stimulating metabolic responses involved in germination.

Response of seeds to light is modified by various internal and external factors. Osmotic stress, the presence of growth promoters or inhibitors, the oxygen tension and several other factors influence the response of seeds to the light.

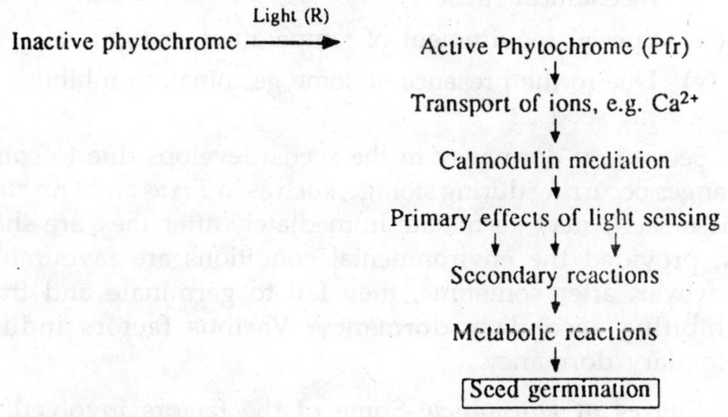

Phytochrome mediated induction of seed germination

5. *Soil Conditions:* The germination of seed in its natural habitat is influenced by soil conditions also, which include water holding capacity, mineral composition and aeration of the soil. Salinity of soil is often an important factor determining the germination and initial seedling growth. The high salt content of soils, inhibit germination due to osmotic effects.

Dormancy of Seeds

Many seeds fail to germinate even if the environmental conditions (water, light, temperature, etc.) are favourable for germination. However, these seeds can be induced to germinate by some suitable treatment of the seeds. This failure to germinate in favourable conditions is the dormancy.

Primary and Secondary Dormancy: Dormancy in seeds can be due to various reasons. When shed off, many seeds fail to germinate because of some intrinsic factor related to the seeds. This is called primary dormancy. This may be due to one of the following limitations;

(i) Immaturity of the embryo,

(ii) Impermeability of the seed coat to water and O_2,

(iii) Restriction in embryo development due to some mechanical cause,

(iv) Special requirement of temperature or light,

(v) Due to the presence of some germination inhibitor in the seed.

Secondary dormancy in the seeds develops due to some changes occurring during storage, such as in *Taxus* and *Fraxinus*. These seeds may germinate immediately after they are shed off, provided the environmental conditions are favourable. However, after sometime, they fail to germinate and thus exhibiting secondary dormancy. Various factors induce secondary dormancy.

Causes of Dormancy: Some of the factors involved in primary and secondary dormancy are described in the following paragraphs.

1. *Permeability of Seed Coat:* Many seeds have a hard and thick seed coat, which is impermeable to water and O_2 and it also restricts the emergence of roots and shoots from-the seed. The seeds of many leguminous plants are hard, resistant to abrasion and covered with a wax like layer. In some seeds, the strophilar cleft, through which water enters inside the seed, is suberised. The area of hilum and chalza are also blocked by special cells.
2. *Immaturity of the Embryo:* Many seeds show primary dormancy because the embryos are immature, when the seeds are shed off. Such seeds have to be stored under suitable environmental conditions, before they can be germinated. Plants in which immature embryos occur, include members from orchidaceae and in *Ranunculus* spp. The period required for such embryo to reach maturity varies from a few days to several months. During this after ripening period several morphological and anatomical changes take place. Some chemical changes inside seed may also take place.
3. *Germination Inhibitors:* A large number of substances present either inside the seed or in its environment inhibit seed germination. The osmotic concentration of the medium in which the seed is germinating is the most important factor controlling the rate of germination. Many seeds fail to germinate when salt concentration in the environment increases by more than 0.5 mM. General inhibitors such as cyanide, dinitrophenol, azide, fluoride, hydroxylamine, etc. inhibit germination in many seeds. However, cyanide at low concentration is known to induce germination in some cases. Herbicides growth retardants (cycocel, phosphon D, AMO 1618, etc.) and phenolic compounds also inhibit germination.

Coumarin (an unsaturated lactone) is of widespread occurrence in many seeds and is a strong germination inhibitor.

It is considered to be natural germination inhibitor, as is the case with abscisic acid. M. Evenari (1949) in a review article has listed a variety of chemical compounds, which are believed to be natural inhibitors in many seeds and fruits. Among these are Phenols, alkaloids, essential oils, aldehydes, parasorbic acid, unsaturated lactones, organic acids and ammonia releasing substances. Among various Phenolics, abscisic acid is a well known germination inhibitor. The degree of dormancy in *Fraxinus americana* and in many other species is well correlated with the abscisic acid content.

4. *Temperature Requirements:* In order to germinate, many seeds require an exposure to some definite temperature prior to being placed at the temperature which is favourable to germination. This treatment involves exposure of fully imbibed seeds to either high or low temperature, at which germination does not occur. Seeds of many rosaceous species, deciduous trees, several conifers and several herbaceous *Polygonum* species fail to germinate unless they are exposed to low temperature in the presence of O_2 in moist conditions for weeks to months. Several other seeds germinate when temperatures alternate between high and low levels.

5. *Light Requirements*: Light generally in combination of several other factors influences germination. Many seeds fail to germinate in dark. However, some seeds germinate in dark only. Temperature, storage period, presence of other chemicals, etc. influence response of seeds to light.

Breaking of Dormancy

Dormancy can be removed and seeds can be initiated to germinate, if the factors causing dormancy are removed. Some of the common methods for breaking dormancy are as follows :

1. *Impaction and Scarification:* As mentioned earlier, the strophiolar opening in some seeds is suberised,

preventing the entry of water and O_2 inside the seed. Vigourous shaking of the seed sometimes removes the suberized plug and allows germination of the seed. This treatment is called impaction and has been applied successfully in the seeds of *Crotallaria, Melilotus, Trigonella,* etc.

Dormancy due to general impermeability of the seed coat can be overcome by partial or total removal of the seed coat. Breaking the seed coat barrier is called *scarification.* Knives, files, and sand paper have been used. Shaking seeds with some abrasives such as sand, also removes the seed coat. Chemical methods are also used for scarification. Chemicals such as alcohol or other solvents dissolve away the waxy layer of seed coat and make it permeable. Alcohol treatment is specially affective for the seeds of the members of caesalpiniaceae. Other chemicals such as sulfuric acid, dissolve the seed coat and bring out some chemical changes as well. Dormancy of cotton seeds is easily broken by soaking the seeds for a few minutes to an hour in cone, sulfuric aid and then washing to remove the acid.

Scarification is also achieved by some natural means, which include microbial action, passage of the seed through the digestive tract of a bird or other animal, or movement by water across sand or rock.

2. *Stratification:* The most commonly known and used procedure for overcoming the dormancy due to a special temperature requirements, is the treatment of seeds at low temperature under moist conditions. This treatment is called stratification. During stratification anatomical and biochemical changes take place inside the seed. The balance in growth substance is also changed during stratification. In *Acer plantoides,* for example, there is a decrease in acidic inhibitors and increase in GA during stratification. In some cases, such as in *Betula pubescens* seeds, stratification can be replaced by treatment of seeds with GA.

3. *Removal of the Inhibitor*: As described earlier, many seeds fail to germinate because of the presence of certain chemical inhibitors inside them. These seeds can be made to germinate by removing these chemicals. Repeated washing with water is an efficient method of removing the" natural inhibitors. Stratification also decreases the inhibitor content in some cases.
4. *Treatment with Growth Hormones:* Germination is greatly influenced by the endogenous and exogenous growth hormones. In many cases, endogenous gibberellin and cytokinin levels are correlated to the germination efficiency of seeds. Gibberellins substitute the cold treatment requirement in many seeds. Cytokinins and ethylene promote germination of many seeds.

Bud Dormancy

Buds of many types of woody plants also undergo dormancy and they fail to grow and develop even under favourable environmental conditions. The dormancy can be overcome, and in some cases, by almost same treatments which are employed for breaking seed dormancy. Some examples are :

(1) Chilling for several weeks at 0-5°C (stratification) is effective in breaking dormancy of buds, rhizomes, corms and many types of seeds.

(2) Certain tree buds and certain seeds may be induced to grow by exposure to long days, where as short days are ineffective.

(3) Certain substances will break the dormancy of several kinds of organs. Thiourea and GA remove the dormancy of buds, rhizomes, corms and many types of seeds.

OTHER ASPECTS

Numerical Methods

Although attempts to use of some kind of quantitative methods in taxonomy date back to the end of nineteenth century, the last 25 years have witnessed marked change in outlook and methodology in the field of biological systematics with the application of high-speed computers. There has been a considerable development of numerical methods which provided numerous new concepts and techniques to systematist. The period from 1957 to 1961 saw the development of first methods and of theory of numerical taxonomy. The publication of pioneer work of Sokal and Sneath (1963) *Principles of Numerical Taxonomy* was the first comprehensive exposition to this subject. Since then there has been a very rapid increase in the development of methods of numerical taxonomy and their application. This field of study is still in a period of active growth as is evident from a large number of papers published every year.

The earlier studies were largely on micro-organisms such as bacteria. But subsequently many studies in widely varying groups of plants and animals, based on numerical methods have appeared and thus 'numerical taxonomy has come of age". There are several publications on applications of numerical taxonomy to angiosperm systematics. Some important ones are on *Apocynum* (Apocynaceae), *Cucurbita* and hybrids (Cucurbitaceae), *Crotalaria* (Fabaceae), *Salix* (Salicaceae), Lamiaceae, Verbenaceae and allied families, *Zinnia* (Asteraceae), *Silene* (Caryophyllaceae), *Chenopodium* and *Atriplex* (Chenopodiaceae), Caucaulidae (Apiaceae), *Quercus* (Fagaceae), *Manihot* (Euphorbiaceae), *Oenothera* (Onagraceae), *Solanum* (Solanaceae), wheat cultivars, Poaceae, Bromeliaceae, maize cultivars, Aloineae (Liliaceae), Onicidiinae (Orchidaceae), and barley cultivars and hybrids (for references see Sneath and Sokal, 1973).

Heywood (1967) defines numerical taxanomy as "the numerical evaluation of the similarity between groups of organisms and the ordering of these groups into higher-ranking taxa on the basis of these similarities." Pointing to the aims of numerical taxonomy Sneath and Sokal (1973) write that "Numerical taxonomy aims to develop methods that are objective, explicit and repeatable, both in evaluation of taxonomic relationships and in the erection of taxa."

Thus numerical taxonomy is based on phenetic evidence as judged from an organism's phenotype rather than from its supposed phylogeny. Numerical taxonomy is divided into a series of repeatable steps and in the following pages a simplified account of the logical steps is given.

Operational Taxonomic Units

The basic units (lowest ranking taxa) in any particular study are normally called Operational Taxonomic Units (or OTU's). These units could be individual or a supra-individual category such as a species, a genus or even a higher ranked taxon. The studies where individuals are employed as basic units would throw light on resemblances among intraspecific variants but would be of little help for comparisons at the subgeneric, generic, and higher levels. When OTU's are species, they could be used for erecting higher taxa. Thus the OTU's may differ in rank from study to study.

Selection of Characters (Attributes): A sufficiently large number (n) of suitable characters are selected of the OTU's. To obtain a fairly stable and reliable classification, it is usually recommended not less than 50 characters should be used and where feasible considerably more characters should be employed. The value of similarity coefficient becomes more stable with the increase in the number of characters sampled. On the other hand, if the number of characters is too small, the statistical precision of coefficients of similarity may also be misleadingly low.

Sneath and Sokal (1973) suggest the following guidelines for the search of taxonomic characters. They should be from all parts and all the stages of the life cycle, and all characters varying within the group studied should be used. Taxonomic characters may be grouped into morphological, physiological and chemical, behavioural and ecological and distributional characters. There are several basic types of attribute data but generally taxonomic studies are characterized by characters (attributes) capable of expression as binary, qualitative multistate or quantitative multistate data.

Binary Characters: They possess two contrasting states such as the presence or absence of some feature being one (*e.g.*, presence or absence of spines) or the same feature being the other (*e.g.*, fruits being dehiscent or indehiscent).

Qualitative Multistate Characters: These characters, also known as disordered multistate characters, possess three or more contrasting forms each ranking equal, *e.g.*, flower colour red, yellow or white. They can be analysed in two ways. The first method of analysis is to convert them to a series of binaries, *e.g.*, flower colours red, yellow or white can be resolved into three sequential presence absence alternatives, *i.e.*, red and not red, the latter leading to yellow and not yellow (this also includes white which is not yellow). In the second method each entity is offered a choice of one of n scores (n= number of states assumed by the attribute). The degree of similarity is measured here in terms of dissimilarity, *i.e.*, less dissimilar entities are more similar.

Quantitative Multistate Characters: They represent measures of size of a continuous scale such as weight or length. Examples are the amount of a chemical produced by a bacterial strain or length of spore. They can be recorded into several two-state characters or arbitrarily dividing the scale into two parts which need not necessarily be equally long.

In numerical taxonomy, the characters or attributes used have to be broken down into unit characters (a character of two or more states which can not be subdivided further). In most of the earlier studies on numerical classification, a single type of attributes, that is binary attribute, has been used. However, in recent years there are many studies where the attributes are not of the same type but usually a mixture of more than one type.

Numerical taxonomists are generally in agreement that all the characters selected be given equal weight when creating taxonomic\ groups. However, there are some workers who do not consider that equal weightage is desirable and they suggested that characters should be weighed depending on the "rank" of the organ which they describe. Arguing in favour of equal weightage of all characters, Sneath and Sokal (1973) state that "it is only practical solution, it and only it can give the sort of natural taxonomy that we want, and it will appear automatically during the mathematical manipulation".

Data Processing

To estimate the similarity between pairs of OTU's the basic data are recorded in an n × t matrix consisting of t OTU's classified by n characters (Table).

Table: Coded data table, t denotes number of OTU's and n chosen number of characters for the study 1, attribute present; 0, attribute absent; NC, no comparison with this entry.

Characters (n)	*OTU's (t)*			
	A	B	C	D
1	1	1	0	NC
2	1	1	1	1
3	1	1	1	1
4	0	1	NC	NC
5	1	1	1	1
6	1	1	0	1
7	1	1	0	NC
8	NG	0	1	1
9	1	1	1	1
10	1	1	1	0

There are various ways to express the scores in a data matrix. In the simplest form of coding of two-state characters (all or none characters) an attribute is divided into two states, "present" or "absent". The + symbol indicates the presence of attribute and the — symbol its absence. However, some workers prefer the use of 1 and 0 to facilitate the numerical analysis. Quantitative multistate characters are divided into several states, 1, 2, 3, 4, etc. and they can be scored for example as continuous variables, rank orders or percentage of maximum expression. Qualitative multistate characters which can not be, ordered are usually illustrated by a series, *,□, §. These may be scored as numbers or alphabetic symbols depending on the computer programme. For missing entries in the data matrix, the symbol NC (no comparison with this entry) is commonly used.

Estimation of Similarity

The resemblance between a pair of OTU's can be expressed in terms of similarity, *i.e.*, the percentage of characters in which they agree or dissimilarity, *i.e.*, the percentage of characters in which they disagree. Since dissimilarity can be expressed as the taxonomic distance between the positions of the OTU's in the phenetic hyperspace, it is particularly useful in numerical taxonomy for the construction of taxonomic maps or models. How similarities or dissimilarities are estimated ? The following example may illustrate the point where two OTU's are scored for the five characters:

Characters	1	2	3	4	5
OTU_i	–	+	+	–	+
OTU_k	+	+	+	–	–

They agree in second, third and fourth and disagree in first and fifth and thus their similarity is 3/5 = 60 per cent and dissimilarity is 2/5 = 40 per cent.

There are several classes of similarity coefficients that had been suggested for biological taxonomy. They are used with different kinds of characters. They are based on comparing the two sets of character data for any given pair of OTU's in terms of the numbers of agreement or disagreements in the characters. The character states of a pair of OTU's j and k are compared over the n characters. The missing observation (denoted by symbol NC) is not considered and 1 is subtracted from n.

In the most common model for qualitative characters, the number of the four possible different combinations of matches or agreements and mismatches or disagreements of the character states of two OTU's can be shown in a conventional 2 × 2 frequency table (Table).

Table: 2 × 2 frequency table.

		OTUj 1	OTUj 0	
OTU_k	1	a	b	a+b
	0	c	d	c+d
		a+c	b+d	n=a+b+c+d

a, b, c and d are the numbers of the four combinations and total of four combinations a + b + c + d = n. If number of matches, m = a + d and number of mismatches, u = b + c, m + u = n. The following table gives a numerical example with 20 characters.

Table: 2 × 2 frequency table with 20 characters.

		OTUj 1	OTUj 0	
OTU_k	1	12	3	15
	0	1	4	5
		13	7	20

The Simple Matching Coefficient (S_{sm})

$$S_{sm} = \frac{m}{m+u} = \frac{m}{n} = \frac{a+d}{a+b+c+d}$$

The example gives $S_{sm} = \frac{(12+4)}{(12+3+1+4)} = 0.80$

The Jacard Coefficient (S_j)

$$S_j = \frac{a}{a+u} = \frac{a}{a+b+c}$$

Negative matches are excluded from consideration in this similarity coefficient.

The example gives $S_j = \frac{12}{12+3+1} = 0.75$

The following are the most commonly used coefficients for quantitative characters:

Taxonomic Distance (d)

$$d_{jk} = \sqrt{\left\{\sum^{n}\left(x_{ij} - Y_{ik}\right)\frac{2}{n}\right\}}$$

where the character state of *OTUj* for character *i* is X_{ij} and that of OTU_k is X_{ik}. The symbol $\sum n$ indicates the sum over n characters. The value of d is the distance in a phenetic space divided by $\sqrt{n}$. The distance with 1, 0, characters is related to S_{sm} as indicated below:

$$d = \sqrt{(1-S_{sm})}$$

The Correlation Coefficient (r)

$$r_{jk} = \frac{\sum^{n}[(X_{ij} - \overline{X}_i)(X_{ik} - \overline{X}_k)]}{\sqrt{\{[\sum^{n}(X_{ij} - \overline{X}_j)^2][\sum^{n}(X_{ik} - X_k)^2]\}}}$$

The symbols $\overline{X}_j$ and X_k are the average values of X over all n characters for OTU's j and k respectively.

Gower (1971) has proposed a general coefficient of similarity applicable to mixed qualitative and quantitative characters.

Gower Coefficient (SG)

$$SG = \sum^{n}\frac{sijk}{n}$$

where sijk is a score on character i for comparison of OTU's j and k such that sijk = l – ($1X_{ij} - X_{ik}1/R_i$). Here R_i is the range of character i over all t OTU's. If all characters are qualitative (1, 0) then SG becomes S_{sm}.

Taxonomic Structure

The similarity matrix (resemblance matrix) is the basic evidence available for finding taxonomic structure. In this t × t matrix of computed similarities, all similarities are arranged to fall on a scale between 0 and 1. Since the similarity of j and k is the same as that of k and j, only one half of this square matrix is filled in. A typical 10 × 10 similarity matrix for 10 OTU's is given in Table.

Table: 10 × 10 similarity matrix for 10 OTU's (after Bailey, 1970).

OTU's	1	2	3	4	5	6	7	8	9	10
1.	1.0									
2.	0.9	1.0								
3.	0.6	0.5	1.0							
4.	0.5	0.6	0.5	1.0						
5.	0.9	0.8	0.6	0.5	1.0					
6.	0.6	0.6	0.8	0.5	0.6	1.0				
7.	0.6	0.6	0.8	0.5	0.5	0.8	1.0			
8.	0.5	0.5	0.5	0.9	0.5	0.5	1.5	1.0		
9.	0.8	0.9	0.6	0.5	0.9	0.6	0.6	0.6	1.0	
10.	0.6	0.6	0.7	0.6	0.6	0.8	0.8	0.5	0.6	1.0

OTU's

Afterwards the OTU's are rearranged to produce a sortex matrix where OIU's with high mutual similarity are arranged together to form phenetic groups.

Table: Rearranged version of similarity matrix in Table. (after Bailey, 1970).

OTU's	1	2	5	9	3	6	1.0	2.0	4	8
1.	1.0									
2.	0.9	1.0								
5.	0.9	0.8	1.0							
9.	0.8	0.9	0.9	1.0						
3.	0.6	0.5	0.6	0.6	1.0					
6.	0.6	0.6	0.6	0.6	0.8	1.0				
7.	0.6	0.6	0.5	0.6	0.8	0.8	1.0			
10.	0.6	0.6	0.6	0.6	0.7	0.8	0.7	1.0		
4.	0.5	0.6	0.5	0.5	0.5	0.5	0.5	0.6	1.0	
8.	0.5	0.5	0.5	0.6	0.5	0.5	0.5	0.5	0.9	1.0

OTU's

On the basis of highest mutual similarity 10 OTU's can be arranged into three discernible clusters (nos. 1, 2, 5 and 9, nos.

3, 6, 7 and 10, and nos. 4 and 8). These three clusters can be regarded as three phenons.

This process of rearrangement is highly subjective as it is based on visual inspection of the similarity matrix. Furthermore, it is difficult to achieve with more than 10 OTU's. An essentially identical method of shaded similarity matrices was described by Sokal and Sneath (1963). Here a range of similarity coefficients is indicated by squares with different densities of shading. This method is commonly used in microbiology.

One of the simplest method of cluster analysis is the single linkage method. In this method the OTU's that show highest similarity level are joined into pairs, and then the OTU's most similar to these pairs are added on at successively lower level of similarity. This process continues until all the units have joined together (Table).

The results of cluster analysis can be represented in a tree-diagram or dendrogram. A tree-diagram based on phenetic evidence is also known as a phenogram. It shows at what similarity levels various clustering occur. The OTU's nos. 1 and 2 show the highest association with a coefficient of 0.9. Next, no. 3 joining the cluster of nos. 1 and 2 at a level of 0.8. The OTU's nos. 5 and 6 form a second cluster at a level of 0.7. The cluster with nos. 1, 2 and 3 is combined with the cluster formed of nos. 5 and 6 at a similarity of 0.45. After this, OTU no. 4

Table: Single linkage clustering. The link between OTU's is in falling order of similarity (after Sneath, 1978).

	OTU's
Similarity	*1 2 3 5 6 4*
1.0	
0.9	1 — 2
	1 — 2
0.8	1 — 2 — 3
	1 — 2 — 3
0.7	1 — 2 — 3 — 5 6

Contd...

Similarity	*OTU's*					
	1	*2*	*3*	*5*	*6*	*4*
	1 — 2 — 3 — 5 6					
0.6	1 — 2 — 3 — 5 6					
	1 — 2 — 3 — 5 6					
0.5	1 — 2 — 3 — 5 6					
	1 — 2 — 3 — 5 — 6					
0.4	1 — 2 — 3 — 5 — 6					
	1 — 2 — 3 — 5 — 6 — 4					
0.3	1 — 2 — 3 — 5 — 6 — 4					

is combined with the composite cluster formed of nos. 1,2,3, 5, and 6 at a similarity level of 0.35 (Fig.). This produces a hierarchical tree in which various clusters define phenons (calculated taxa).

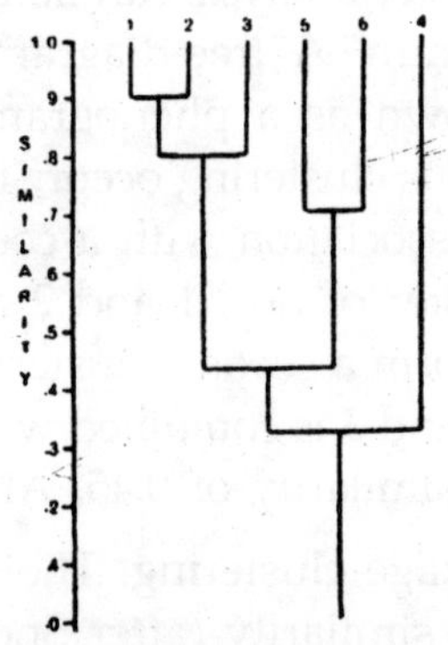

The single linkage phenogram derived from Table (after Sneath, 1978).

There are several other procedures for representing taxonomic structures, most of these are ancillary to phenograms. Some three-dimensional models have also been proposed.

In numerical taxonomy all the powerful techniques of mathematics and computer science are employed to provide a method of taxonomic analysis with logical basis.

The methods of numerical taxonomy have been successfully employed to a wide range of biological organisms, and to a great extent they provide results that agree with established classification. Thus they are highly relevant to the real problem of biological classification. Being mathematical and quantitative, numerical taxonomy provides a more exact basis for classification and it also permits the classification to be built on a wide range of characters. Numerical taxonomy is also proving helpful in the study of phylogeny and in the preparation of taxonomic keys.

3

Growth of Vegetables

Growth and development are the most fundamental and conspicuous characteristics of all living organisms. According to dictionary, growth is "the advancement towards or attainment of full size or maturity" and development; "a gradual increase in size." Almost similar criteria apply for plant growth as well. Accordingly, the plant physiological definition of growth is—*an irreversible increase in mass, weight or volume of living organism, organ or cell.* The key phrase in the definition is 'irreversible increase'.

The increase in any of the parameters (mass, weight or volume) must be irreversible or permanent. Thus, swelling of a cell or organ is not growth, as the cell or organ can be brought to its original size by returning it to a concentrated solution. It has also to be realized that only one (or more) of the parameters has to increase permanently.

In certain kinds of growth, while one parameter may increase, the other may remain unchanged or even decrease. For example, during seed germination and early seedling growth (specially in the dark), while there is an increase in the volume or length and fresh weight of the seedling, there is a net decrease in dry weight. Nevertheless, seed germination and seedling formation are examples of growth.

Growth leads to the development. Development is defined as "ordered change or progress, often (but not always) towards a higher, more ordered, or more complex state" (R.G.S. Bidwell). According to this definition, development can take place without growth, and growth without development. However, as mentioned earlier, the two processes are often linked together and occur in sequence.

Another terminology, used in conjunction with the growth and development is the differentiation. Differentiation is the process, by which cells become specialized. For example, the product of mitotic cell division is an undifferentiated cell. But the cell, later on is modified to produce various kinds of tissues, such as xylem, phloem, etc. This modification of the cell to produce a specialized tissue is differentiation.

The growth, development and differentiation are often linked processes, in that order.

Measurement of Growth

Growth can be measured by a variety of parameters. Some of these are as follows:

1. *Fresh Weight:* Determination of fresh weight is an easy and convenient method of measuring growth. In some cases, as for a potted plant, the measurement can be taken without any injury or damage to the plant.

   For measuring fresh weight, the entire plant or desired organ (root, shoot, leaf, etc.) of the plant is harvested, cleaned for dirt particles, if any; blotted dry for any water sticking on the surface and then weighed.

2. *Dry Weight:* The dry weight of the plant organ is usually obtained by drying the material for 24 to 48 h at 70 to 80°C and then weighing it. Alternatively, the sample can be freeze dried. The measurements of dry weight may give a more valid and meaningful estimate of

growth than fresh weight. It also gives much more consistent (within the replicates) value than the fresh weight. However as stated earlier, in measuring the growth of a dark grown seedling, it is desirable that the fresh weight is determined.

3. *Length*: Measurement of length is a suitable indication of growth for those organs, which grow in one direction with almost uniform diameter, such as roots, shoots or pollen tubes. The length can be measured by a scale or with a calipers if precision is required. Graph papers can also be used for measuring growth of roots and shoots. In case of pollen tubes, the length is measured by a micrometer device fitted in a microscope.

The advantage of measuring length is that it can be done on the same organ over a period of time, without destroying it.

4. *Area:* It is used for measuring growth of those organs which grow mainly in two directions, such as leaf. The area can be measured by a graph paper or by a suitable mechanical device. The most modern laboratories, these days, use a photoelectric device, which reads leaf area directly as the individual leaves are fed into it.

Another common method of measuring leaf area is using linear regression analysis;

$$\text{Area} = ab\ (1 \times w)$$

where a is a constant, b is slope, 1 is length of the leaf and w is the width of the leaf.

For bean plants, the values of a and b determined from 60 leaves have been found to be 0.624 and 0.583, respectively.

Growth Rates—Kinetics of Growth: The change in size or weight of an organism or a organ is often measured as a function of time. The measurement can be made from two view points *absolute growth rate* and *relative growth rate*. Absolute growth rate (AGR) is total growth of each plant or organ per unit time. It is used when the yield of a plant or plant organ is to be determined. Relative growth rate (RGR) is the growth of each per unit time expressed per unit of the critical weight of volume. If the growth activities of two plants or plant organs are to be compared, then RGR is to be measured. The difference between AGR and RGR can be explained further by following example: If two leaves with an initial area of 5 and 10 cm^2 have increased to 50 and 60 cm^2 respectively to 5 days, then the AGR for both leaves will be the same, that is 10 cm^2 per day. However, the RGR of both leaves will be different, because their initial areas were different. Thus, in 5 days, while smaller leaf has increased 11 times (from 5 to 55 cm^2), the larger leaf has increased only 6 times (from 10 to 60). The RGR of smaller leaf is therefore about 1.8 times (11 divided by 6) higher than that of the larger leaf.

In terms of mathematical equation, the AGR and RGR can be expressed with following equations:

$$\text{Absolute growth rate (G)} = \frac{dw}{dt}$$

where dw is increase in weight and dt is change in time.

or
$$G = \frac{w_2 - w_1}{t_2 - t_1}$$

where w_2 is the final weight and W_1 is initial weight (or area), t_2 is final time and t_1 is initial time.

$$\text{Relative growth rate (R)} = \frac{1}{w} \cdot \frac{dw}{dt}$$

or
$$R = \frac{1}{w_1} \cdot \frac{w_2 - w_1}{t_2 - t_1}$$

The symbols are the same as in the equation for absolute growth rate. Relative growth rate is also called specific growth rate.

A plant or a plant organ follows a regular growth pattern during its life. A typical growth pattern of an annual plant or of individual parts of both annual and perennial plants is shown in figure. It is sigmoid (s-shaped) and can be divided into three phases: (i) a logarithmic phase, (ii) a linear phase and (iii) a phase of declining rate or senescence phase. In the logarithmic phase, the growth rate (the increase in area or weight per unit time) is slow at first, but the rate continuously increases after sometime. A logarithmic phase is also exhibited by cultures of bacterial, and other single called organisms. In the linear phase, increase in size or weight continues at a constant, usually maximum rate for some time. The senescence phase I characterised by a decreasing growth rate, as the plant reaches maturity an begins to sense.

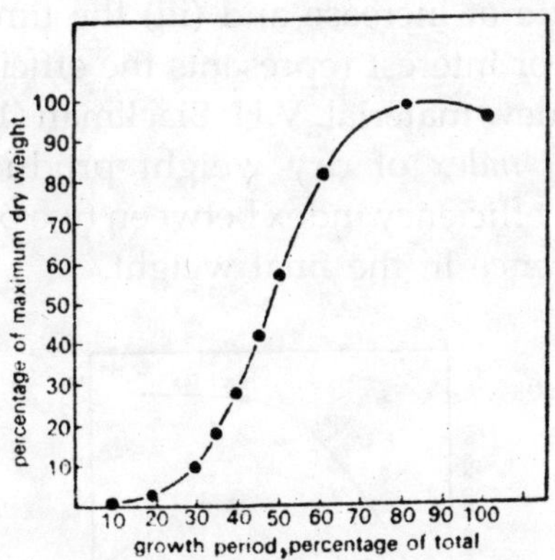

Increase in dry weight of barley Plants during growing season.

The growth curve can be subjected to mathematical treatment. The mode or formula for each part of the growth curve can be constructed and can be used to test the effect of an unknown factor on growth. The mathematical analysis of the log phase of the growth is rather simple. If each cell or fixed population of cell in an organism divides into two at regular intervals and daughter cells grow to the same size as original one, then the organism increases by compound interest law; -

$$\text{Relative growth rate (R)} = 2.303 \log \frac{\text{Increase in mass}}{\text{Time interval}}$$

or $$R = 2.303 \log \frac{\text{(Final mass- initial mass)}}{\text{(Final time - initial time)}}$$

By employing the equation, the final mass of an organism can be calculated if the rate of growth is known;

Final mass = Initial mass$^{er \text{ (time interval)}}$

where e is the base of natural logarithm (2.303) and r is the growth rate.

This equation is the form of y = a + bx. This means that we should obtain a straight line, when we plot a log mass (weight) against time. This has been demonstrated in a number of cases. From the equation, it is also clear that the final mass of the organism will depend upon: (i) the initial weight or mass, (ii) the rate of increase and (iii) the time interval. The rate of increase or interest represents the efficiency of a plant as a produce of new material. V.H. Blackman (1919) called this as the *efficiency index* of dry weight production. A small difference in the efficiency index between two plants will make a marked difference in the final weight.

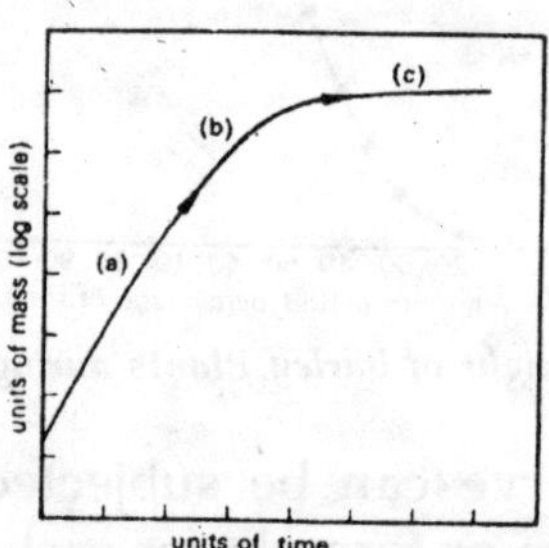

Logarithmic plot of the growth curve in figure

(a) exponential phase, (b) linear phase, (c) senescence phase.

The linear shape of the logarithmic phase *i.e.*, growth plotted on the logarithmic scale is seen in cultures of growing single cell organisms, such as bacteria, or in the growth of mass cells growing in every direction. Most higher plants develop a meristematic growth plan, after this initial phase of growth. Growth takes place only at certain places (meristems). Growth of meristem is essentially one dimensional, that is an increase in length. This phase of growth, the linear phase, is not related to the size of the organism. The mathematical expression for this phase of growth is;

$$\text{Growth rate} \quad (R) = \frac{\text{Increase in mass}}{\text{Time interval}}$$

$$(R) = \frac{(\text{Final mass - initial mass})}{(\text{Final time - initial time})}$$

By employing this equation, the final mass of an organism can be calculated, if the rate of growth is known;

Final mass = Initial mass + growth rate (time interval).

Features of Plant Growth

The growth curve described in the earlier section is seen in most cases, although there is considerable variation due to plant, as well as environmental factors. It is also apparent that, growth in all parts of a multicellular plant is not uniform. In higher plants, it is restricted only to the meristems. The principal meristematic zones are found near the root and the shoot tips, in the vascular cambium, near the nodes of monocot and in certain parts of young leaves.

The root and shoot, apices are active in growth, even during the seed germination, while cambium and leaf meristems are functional only after the seedling is established. Based on the location and activity of meristems, some terminologies are used to define the growth and development of an organ or a plant. Some of these terms are described in the following paragraphs.

Determinate and Indeterminate Organs: Those organs which grow to a certain size and then stop growing, eventually going senescence and death, are called determinate organs. Example of such organs are leaves, flowers, fruits, etc.

Those organs or structures, which grow on continuously with the activity of meristem, are indeterminate organs.

Examples are roots and vegetative stems of perennials. These structures always remain youthful, because of the meristems.

Monocarpic and Polycarpic Species: Monocarpic species flower only once and then die. Thus, in a sense, monocarpic species are determinate, as far as the entire plant is concerned.

Polycarpic species flower more than once and there is a vegetative period between two periods of flowering. Thus, these are viewed as indeterminate plants.

Most monocarpic species are annuals. However, some of them are biennials and perennials also. Many varieties of bamboos (*Bambusa* spp.) may grow and live for over 50 years and then they flower and die. Thus bamboos are perennial but monocarpic. All polycarpic plants are perennials. They use come of their buds for the formation of flowers and other buds develop into vegetative organs.

Range of Growth: Growth rates of plants vary according to the species, age, nutritional and various environmental factors. Under optimum conditions, some species have a very fast growth, in the range of 1 to 4 cm a day, while others may grow only a few cm in a year. Examples of some fast growing species are maize (13 cm a day), asparagus (30 cm a day), bamboo (60 cm a day), etc. The growth rate of pumpkin fruit is also extremely high. Most perennials have a slow growth rate.

Root Growth

Radicle is the embryonic root. During seed germination and seedling formation, it grows to form primary root of the seedling. It may branch further to produce secondary and tertiary roots. A growing root usually has 4 distinct regions; the root cap, the meristematic region, the region of cell elongation and the region of differentiation and maturation. But in some roots, these regions are not very distinct.

The root cap protects the root tip. The meristematic region in young root is situated just below the root tip. The cells in this region are responsible for growth in the root. The meristematic region typically consists of numerous small, compactly arranged thin walled cells, almost completely filled with cytoplasm. They have very small vacuoles and comparatively large nuclei. Intercellular spaces are absent. Only a few cells in the meristem may actually be involved in the longitudinal growth of the root. F.A.L. Clowes and B.E. Juniper (1968) by employing autoradiographic technique demonstrated that there is a 'quiescent center' in the meristematic region, where no cell division takes place.

This center is located just below the root tip. It is surrounded by a group or layer of actively dividing cells, which give rise to the column of cells forming the root. However, if the meristem is damaged, then the cells of quiescent center become active and form a meristem.

As the root grows and advances in the soil, cells of root cap and tip are torn off by contact with soil particles. These are replaced with new cells formed at the apex of the meristem.

The region of cell elongation is made up of column of newly derived cells. They begin to elongate as soon as the root tip starts advancing. It is the elongation of these cells, which causes the root tip to project forward and push through. Most cells in this region elongate at least 15-folds, while increasing in diameter perhaps only two or three folds. Very considerable pressures are developed by the elongation of root. This can be seen as cracking of old walls or concrete side walks by the roots of the plants inhabiting them.

The cells in the region of differentiation and maturation differentiate into various tissues, characteristic of the mature root; the epidermis, cortex and stele. In roots, both xylem and phloem differentiate only acropetally and as continuation of the older xylem and phloem in the more basal part of the root. In general, phloem tissues differentiate much closer to the apex of the root than xylem tissues. In some root tips, developing sieve tubes are present quite close to the region where cell division is occurring. But the first differentiation of xylem tissues, occurs at least twice as far back of the apical initials as first differentiation of phloem tissues. During differentiation and maturation, most cells increase in size and vacuolation. In sieve tubes (phloem) most of the cell organelles disintegrate during differentiation.

Formation of Lateral Roots

Lateral or branch roots arise from the main root, several mm to a few cm proximal to the root tip. They arise from the pericycle, which is the parenchymatous outer layer of the stele. Prior to the formation of a branch root primordia, the cells of the pericycle in the differentiated region of root, become meristematic. These meristematic cells usually develop opposite the protoxylem points. They divide and grow outwardly through the cortex and epidermis. Possibly some hydrolytic enzymes are involved in this outward growth of the meristem and formation of a root primordia. The enzymes digest the cortical and epidermal cells. The cells elongate and differentiate to form a rootlet. The vascular elements then connect with the vascular elements of the main root, by differentiation of cells behind it.

The plant hormone auxin, is known to promote the formation of lateral roots. Cytokinins inhibit it.

Stem Growth: The life of a stem starts as a plumule. It grows to form the shoot of the seedling. The organization of the meristem in the stem tip, is fundamentally similar to that in the root tip. But its behaviour is a bit complicated because it produces not only the stem but also the branches, leaves, flowers, and other appendages. The longitudinal growth of the stem and formation of various appendages is the function of stem meristem.

The cells in a stem tip meristem are thin walled and approximately isodiametric. They have small vacuoles and larger nuclei. Only a few cells of the meristem are involved in cell division and formation of aerial organs. To explain the cellular organization of stem meristem, A. Schmidt (1924) first proposed a *tunica corpus* theory. Accordingly, most apical meristems contain two zones: an outer tunica and an inner corpus.

Tunica consists of one to several layers of cells at the surface of the meristem. Corpus cells are beneath the tunica layer. The cells in tunica divide by anticlinal division *i.e.,* in a plane perpendicular to the surface of the stem, where as the corpus cells divide in many different planes. However, the division into tunica and corpus layers is not very distinct in many species.

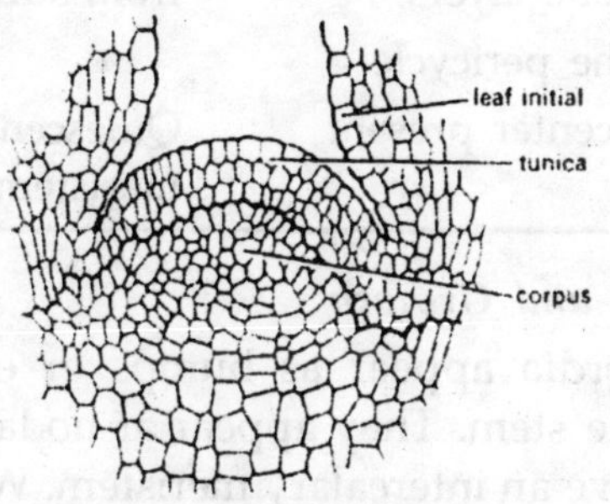

Longitudinal section of vinca rosea apical meristem.

The formation of branches, leaves or other appendages on the stem is initiated by the formation of a primordium or outgrowth at the surface of the meristem, just below the tip. In this respect, shoot growth differs from that of root growth, where lateral root primordia develop from the pericycle. In the formation of aerial organs both tunica as well as corpus layers are involved. Perhaps the corpus enlarges and protrudes first and then tunica development follows. Extensive cell division, elongation and maturation takes place in the two layers. The tunica normally forms the epidermis of the organ derived from the meristem, while corpus cells produce majority of the internal tissues of the new organ.

Auxins normally promote the elongation of stem. They induce the elongation of cells. Gibberellins also promote stem elongation and they do this by promoting cell division as well as cell elongation.

The important differences between root and stem growth are given in table.

Comparison of Root and Stem Formation

Root	Stem
1. Meristem sub apical	Meristem apical
2. Lateral organs arise some distance from apex	Lateral organs arise near apex of the meristem
3. Lateral organs arise from internal tissue layers, normally the pericycle	Lateral organs arise from surface layers
4. Quiescent center present	Quiescent center absent in meristem

Leaf Initiation and Growth

Leaf primordia appear as bumps or elevations on the periphery of the stem. They appear at nodal positions of the stem, which have an intercalary meristem. When leaves are to be produced in pairs, each pair usually appears at right angle to the preceding pair. The two leaves in a pair are generally opposite to each other. Further leaf development is quite variable. The cells of only the primordium tip divide periclinally and anticlinally. Later on, when the leaf is about a few mm or so long, meristematic activity begins throughout its length. The lateral meristems grow more or less rapidly at different positions along the leaf margin, thus giving the characteristic shape of the leaves.

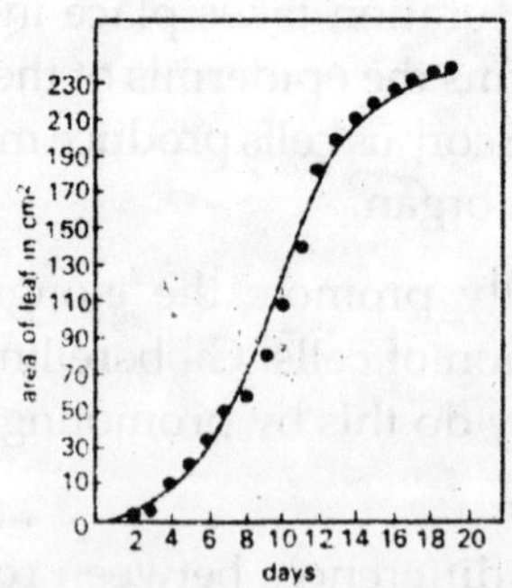

Sigmoidal growth curve of a cucumber leaf.

The growth of an individual leaf also follows the typical sigmoidal pattern, just like the growth of the entire plant. All the three phases of growth, a logarithmic phase, a linear phase and a senescence phase, are apparent.

In most plants, the shape and form of leaves are fixed and little variation is found among them. However, many plants have leaves of different shapes. The phenomenon is termed *heterophylly*. Heterophylly is quite common in aquatic plants. The shape of the submerged leaves is often different from the shape of the floating leaves.

Plastochron and Plant Growth

The knowledge of plastochron is a way of analysing growth of the plant without destroying or disturbing the plant. The term plastochron was first used by the German botanist Eugen Askenasy in 1880 in a description of shoot growth of *Nitella* to indicate the time interval from the formation of one internode cell to the next. Later on the term was used by several plant physiologists to indicate the time interval between the formation of successive leaves on the shoot.

The term plastochron is used sometimes to explain the age of a plant or of a leaf. While studying the growth of *Xanthium* leaves, R.O. Erickson and FJ. Michelini (1957) suggested a reference length of the leaf. It was chosen to be 10 mm. If a plant has 10 successive leaves on it and if its 5th leaf is 10 mm long, then the plant will have an age of 5 plastochron. In other words, the plastochron age of plant is that number which denotes the leaf in succession which is 10mm long.

If the 5th leaf is somewhat longer than 10 mm, then the plant will be somewhat older than the 5 plastochron, but perhaps not yet 6. From the measured length of the 5th leaf and the next youngest one (6th), a fraction of plastochron can be calculated; that added to five will be the plastochron age of the plant.

It should be realized that the plastochron age of a single leaf is not the same as the age of the plant or shoot. A given leaf, which is exactly 10 mm long has a plastochron index of zero, and one longer than 10 mm will have a positive value of the index.

Plastochron Index: The plastochron value of a- shoot can be determined as plastochron index (P.I.) according to the following formulae:

$$P.I. = n + \frac{\log 1n - \log 10}{\log 1n - \log n + 1}$$

where n is the serial number of leaf just longer than the reference length *i.e.,* 10 mm, and log 1n is the logarithm of the length of the leaf 1n.

Plastochron Ratio: Leaves of different ages differ in their length, weight and other attributes. There is almost a constant relationship in length, weight or some other physiological parameters, between leaves of two successive plastochrons. The ratio by which a typical leaf increases in length (or weight) per plastochron is termed as plastochron ratio. Plastochron index is a suitable constant for studying the vegetative growth of herbaceous dicots.

Factors Affecting Growth

Light: Light is an important determinant for many physiological processes, such as photosynthesis, seed germination, flowering, etc. As such it is bound to influence the overall growth of the plant. The visible solar radiation is in the region of 400 to 760 nm and it is absorbed by the plant pigments.

It was rather natural for early botanists to observe the effect of light/dark conditions on plant growth. Perhaps John Ray was the first botanist to mention the physiological effects of light on plants in 1686. C. Bonnet in 1754 carried on some experiments with the beans, peas and branches of vine from which it was recorded that dark grown etiolated plants became green after illumination for 24 hours. Numerous other botanists also examined the effect of light on plants, and by the end of nineteenth century most of the light effects on plant growth and morphogenesis were well documented. Some effects of light on vegetative growth are described in the following paragraphs.

Light Intensity: Most plant grow well only when there is a full sun light. Very low light and very high light cause abnormal growth. Plant grown in total absence of light are etiolated. Their stems have an excessive growth, leaf blade growth is lacking or reduced and there is no chlorophyll formation.

As such the plant looks pale yellow. Plants raised in normal light have reduced stem growth, normal leaf growth with green colour. However, pine needles are green even in the continuous darkness.

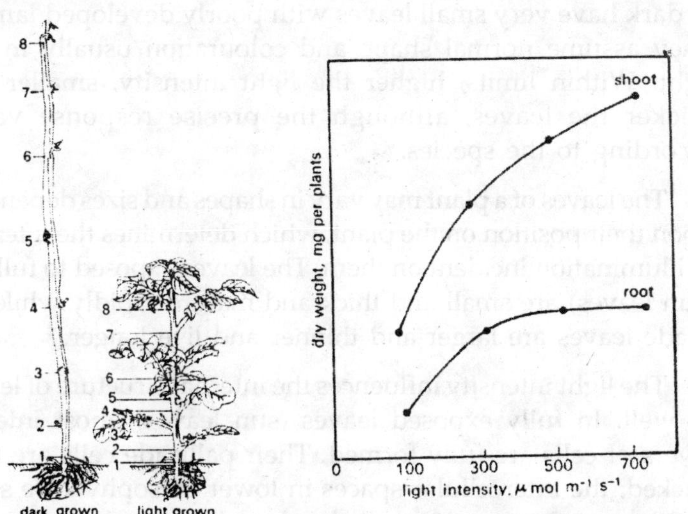

Difference in the growth of potato plants grow in dark and light

Effect of light on increase in weight of bean root and shoot.

The optimum light intensity for various species varies according to the species. The growth of shoot increases with the increase in light intensity, until optimum level. Since roots also depend upon the shoots for carbohydrate supply, their growth is also influenced by light intensity.

Most of the dicotyledons have their hypocotyl in the form of a curved structure or hook during early seedling formation. Once these hooks come above the soil level, they become straight. However, if the plants are kept in dark, the hooks do not open or straighten out. The hook opening is most responsive to red light and the response is negated by far-red light, suggesting the involvement of phytochrome in light perception.

The mechanical tissues in dark or low light grown plants are almost completely absent. The cellular organization is also poor.

Leaves are the organs most commonly involved in light perception and are also themselves greatly influenced by light. Leaf unrolling in grasses takes place only in light. Plants grown in dark have very small leaves with poorly developed lamina. They assume normal shape and colouration usually in low light. Within limits, higher the light intensity, smaller and thicker the leaves, although the precise response varies according to the species.

The leaves of a plant may vary in shapes and sizes depending upon their position on the plant, which determines the intensity of illumination incident on them. The leaves exposed to full sun (sun leaves) are small and thick and mature rapidly while the shade leaves are larger and thinner and live longer.

The light intensity influences the internal structure of leaves as well. In fully exposed leaves (sun leaves), most internal layers of cells are fully formed. Their pallisade cells are fully packed, the intercellular spaces in lower mesophyll are small and vascular system is well developed. In shade leaves, the pallisade tissue is poorly developed and vascular system is almost absent.

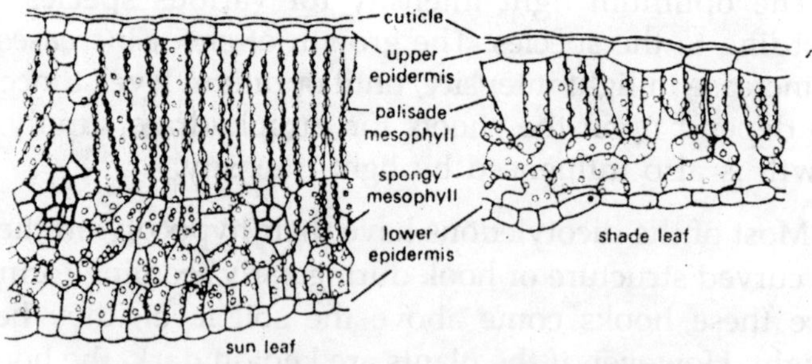

Tranverse section of a sun and a shade leaf showing difference in their tissue organization.

The chlorophyll content and therefore the colour of the leaves is greatly influenced by the light intensity. Generally the optimum light intensity for chlorophyll formation is lower than that for growth. In bean leaf for example, maximum chlorophyll content is at about 80 to 120 μ.mole $m^{-1}s^{-1}$ of continuous illumination, while growth of the shoot increases upto 700 μ mole $m^{-1}s^{-1}$ at least.

High light intensity inhibits chlorophyll formation and accelerates its breakdown. Therefore, the leaves in very high light intensity look pale. Chlorosis in high light is specially apparent in shade loving plants. Some plants protect their chlorophyll loss at high light intensity, by migration of the chloroplasts to the center of the cell.

Light Duration: Many aspects of vegetative growth are modified when plants grow in different day lengths. Some of these are closely linked with the flowering, while other are independent responses. Leaf growth, stem elongation, rooting capacity, branching habit and anthocyanin formation are independently affected by photoperiod. In *Lotus corniculatus,* when plants are grown at 13 h photoperiod, they are semi erect to prostate with short internodes.

However, when they are raised at 14 h photoperiod, they are more or less erect with normal internodes. In some pine species, the shorter day lengths cause premature fading and discolouration of needles. The green colour of the needles reappear when the plants are placed in longer photoperiods. Continuous illumination can cause chlorosis also in some species.

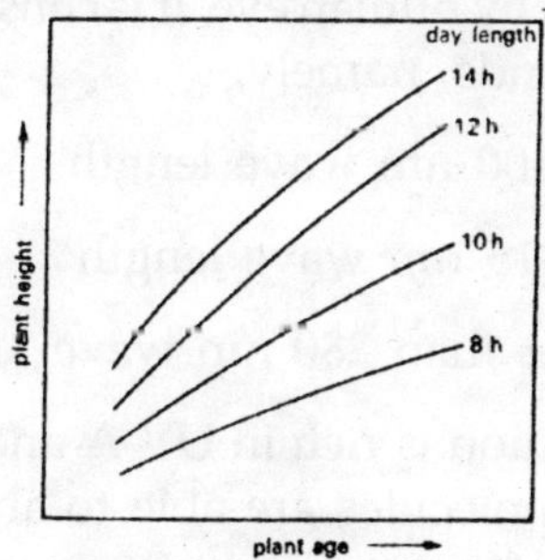

A typical response of plant height to day length.

In most species the photoperiod has a marked influence on stem elongation. Plants grown in long days are generally taller than their counter parts in short days. The stem length increases with the increase in the length of the photoperiod, of course, upto a limit. In *Rudbeckia,* long days initiate blooming as well as the elongation of stems. Under short days, the plants have virtually no internodes and their leaves are close to the ground in a pattern known as rosette.

Light Quality: Most growth and morphogenetic responses are evoked by red light. In many species, it has been found that red light increases stem length, internode length and leaf length while blue light decreases it. In most of the photomorphogenetic effects, the receptor pigment is phytochrome and therefore the effect of light is red-far-red reversible. For example, the elongation of dark grown seedlings is promoted by red light and this effect is reversed if the seedlings are exposed to far red light.

However, for overall growth of the plant, full spectrum of sunlight seems to be necessary. The greatest increase in dry weight usually occurs when full spectrum of sunlight is given. Expansion of leaf blade is prevented by darkness, retarded by green light, intermediate in blue and the greatest in white light.

Ultraviolet Radiations: Solar radiation incident on earth surface and on plant canopy contains ultraviolet (UV) radiations, whose wavelengths are just shorter than those normally perceived by human eye. It is considered to be divided into three basic bands, namely;

UV A → 315 – 400 nm wave length

UV B → 280 – 315 nm wave length

UV C → shorter than 280 nm wave length

The solar radiation is rich in UV A and UV B. The nucleic acid and protein molecules are able to absorb UV radiations and significant effects are seen on plant growth and development.

Under experimental conditions, exposure of plants to UV light has been found to reduce plant growth. Other observed effects of UV are: altered root/shoot ratio, depressed flower development, loss of apical dominance and abscission of leaves.

Biochemical Aspects of Light Effects on Growth: The effect of light on plant growth and development is mediated via its effects on biochemical processes. As far as the growth is concerned, it appears that the most important effect is the change in cell wall properties. The effect, however, does not seem to be direct, but apparently through some other processes. This is demonstrated by the fact that there is a considerable lag period in the effect of light on change in cell wall properties. In blue light, the lag is of about 30 seconds, but in far red responses (the effective light in growth), the lag period is of 10 to 15 minutes.

As compared to this, the effect of light on photochemical and enzymic reactions is very fast, in the order of 10^{-9} to 10^{-3} seconds. Hence, it is possible that light modulates cellular processes that control wall formation.

Several suggestions have been made regarding the effect of light on cellular processes. Some of these are:

(1) Light acts through the synthesis of plant hormones.

(2) It changes the ionic environment of the cell wall.

(3) It alters the rate of synthesis of cell wall components.

(4) It changes the activity of hypothetical wall loosening enzyme.

Temperature: Temperature affects growth and developmental processes in several ways. As far as growth is concerned, it affects both, rate as well as total growth. In the lower range of temperature, the growth increases with the increase in temperature, until it reaches maximum at optimum temperature. The rate of growth decreases with further increase in temperature. It is a typical response of growth to temperature, as is seen in enzyme activity.

In lower range of temperature, the growth rates, like enzyme activity, are more than doubled by a 10°C rise in temperature *i.e.*, Q_{10} is more than 2. But the value of Q_{10} decreases with the increase in temperature range. For example, in maize, Q_{10} between 12 and 22°C is 5 and between 20 and 30°C is 2.5.

Each species has an optimum temperature or range of temperature for growth, although the optimum may vary throughout the life of the plant. The growth of each species is typically adapted to its natural temperature environment. Plants native to warm habitats require higher temperature for growth than those of the cooler regions. The optimum growth temperature for winter wheat, a cereal of cool temperate climate is 20-25°C; for maize, a cereal from warmer climate, it is 30-35°C.

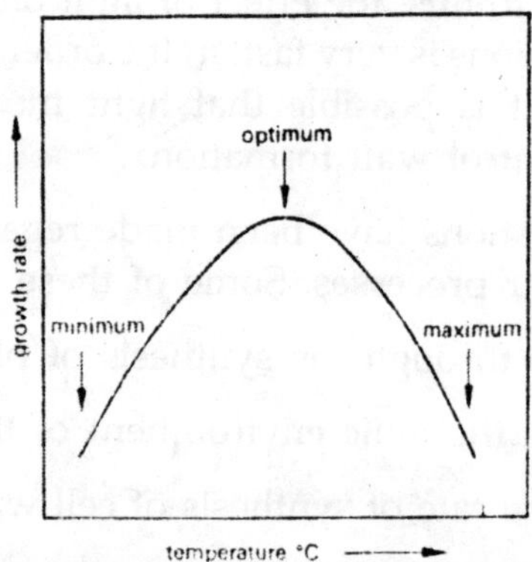

A typical response of plant growth to the temperature.

There is also a minimum temperature, below which growth does not take place. Very low temperature can cause the death of the plant. Above a certain maximum temperature also growth does not take place.

Temperatures higher than the maximum temperature are injurious and can kill the plants. The minimum and maximum limits of the temperature for plant growth varies according to the species (Table).

Temperature Limits for the Growth of Some Plants.

Species	*Temperature °C*	
	Minimum	*Maximum*
Acer plantoides	7-8	26
Cucurbita pepo	14	46
Cucumis sativus	15-18	44-50
Pisum sativum	5-7	44
Sinapis alba	0	37
Triticum vulgare	0-5	42
Zea mays	9	46

In most species, when temperature drops below minimum or increases above the maximum, the plants are usually killed. When temperature falls below freezing, ice crystals may form inside the cell and actually pierce cell walls and membranes. How perennial plants of cold climate tolerate freezing winter is not clearly understood, but it is believed that increasing viscosity of the cytoplasm of the cells, is a major factor in preventing ice crystal formation.

Plant tissues also tend to dehydrate in winter, as they rehydrate in spring, just as growth begins. At higher temperatures, the death of the plant occurs due to inactivation of enzymes and other macro-molecules, essential for life.

Thermoperiodism: The term thermoperiodism was coined by Fritz Went (1957) for promotion of growth by lower night temperature. It may be defined as 'promotion of growth and/or development by alternate day-night temperature'. Went observed that the growth of tomato plants was better when they were grown at 25°C (day) and 17°C (night) temperatures than at a constant temperature.

Formation of fruits in tomato and tubers in potato have also been found to increase with alternating day-night temperatures. Stem elongation and flower initiation are also thermoperiodic responses in some species. Lower night temperature often results in higher sugar content of plants and may also produce greater root growth.

The optimum thermoperiod may change with the growth stage of the plant. Thus, younger pepper plants have been shown to develop best when night temperatures are approximately 25°C, but as they grow older the optimum night temperature for their development drops to 11°C or lower. Not all species show a thermoperiodic response. Cocklebur, sugar beet, wheat, oat, bean and pea grow as well at an optimum constant temperature as they do when day-night temperatures vary.

Soil Water Content

One of the most widely limiting factors for plant production on a global basis is water. Dry areas, whether hot or cold, are devoid of any vegetation and are deserts. On the other hand, the areas which receive abundant rain fall are jungles, even though they may have extreme temperature regimes. Thick forests in equatorial regions and in Tundra are examples of such vegetation. In agricultural practices also, crop yields are often increased by irrigation.

Most plants grow well, only when there is an adequate supply of water in the soil. The growth decreases when the water content of soil decreases beyond a certain critical low level and it also decreases in water saturated soils. The effects of drought conditions (decreased soil water) and water logging are described in the following paragraphs.

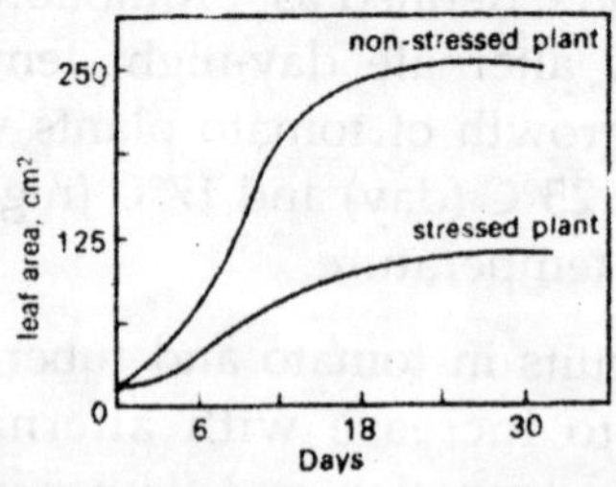

Effect of water stress on leaf expansion growth in Cacao

Drought Conditions: Drought conditions create a water stress in the plant. A temporary water stress in the plant may also be created when rate of transpiration far exceeds over that of water absorption from the soil. Such conditions prevail over during mid-day in hot summer months. Although temporary water stress may cause some physiological disturbances, such as stomatal closure and reduced gas exchange, it may not have any significant effect on overall plant growth, unless it is too frequent.

However, prolonged water stress reduces root and shoot growth, leaf area expansion, overall plant growth and yield. Shoot growth is reduced more than root, as a result root/shoot ratio increases during drought conditions. Leaf area expansion is the most sensitive parameter and it is reduced with even a small water stress, well before many other physiological parameters are affected.

The effects of water stress on cellular and metabolic activities are much more pronounced, which of course are reflected in decreased growth and yield. Decreasing the external water potential by only a bar (unit of water potential) or less, results in a significant decrease in cellular growth. Among different phases of cellular growth, cell enlargement is inhibited more than the cell division or differentiation.

The decreased cellular growth is often accompanied with several metabolic activities, such as decreased protein synthesis, cell wall formation, several enzyme activities, photosynthesis, etc. The growth hormones abscisic acid and ethylene play an important role in water stressed metabolism. Both of these hormones increase in water stressed plants.

Not all species respond to drought conditions in a distinctive manner and many species have adapted to drought conditions. Basically there are three types of drought adaptations in plants : (1) Acquired hardiness, (2) Xerophytism and (3) Dehydration.

(1) Acquired hardiness is the induced resistance for withstanding drought conditions. This phenomenon

was observed as early as in 1883, when H. Will reported that repeated soaking and drying of seeds resulted in increased drought and frost resistance by the vegetatively growing plants. Several reports have been published by others also, and over the years. Some cereals, cabbage, soybean and many others seem to acquire hardiness to drought, if they are exposed to water shortage at seedling stage. Hardiness resulting from either pre-sowing treatment of seeds or of seedlings is manifested by various types of physico-chemical changes in the plant. Some of the probable changes in the cytoplasm are:

(i) Greater hydration of colloids,

(ii) Higher viscosity and elasticity of the cytoplasm,

(iii) Increase in bound water,

(iv) Increase in hydrophilic and decrease in lipophilic colloids and

(v) Increase in temperature required for protein synthesis.

(2) Xerophytism is the morphological adaptation of plants to dry conditions. The leaf area is greatly reduced, so as to avoid transpiration. Some xerophytes develop water storage tissues inside their body.

(3) Dehydration is a natural decrease in water content either during developmental processes or as such. Dry plant tissues can withstand drought conditions. It is well known that seeds actively lose their water content during fruit ripening. Dry seeds develop tolerance to drought. Lichens, mosses and some ferns can also live in dry conditions for prolonged pereopods and can tolerate dehydration.

Water Logging Conditions: Water logging in the soil creates anaerobiosis and it seriously retards the growth of most mesophytic plants. It affects almost each part of the plant.

Because of slow gaseous diffusion in the soil, the root growth is most severely affected by water logging. If the water logged anaerobic condition continues for sometime, roots of most plants do not survive after a few hours. The survival time in anaerobic condition also depends upon species, to some extent. For example, while the roots of *Gossypium hirsutum* die within 0.5 to 3 hours, those of *Scirpus maritima* can survive upto 2 months.

Inhibition of stem growth is also apparent in water logged conditions, but for aquatic plants and marsh dwellers such as *Ranunculus.* In many herbaceous species the base of the stem or hypocotyl gets swollen in water logged conditions. It is a result of cell enlargement in the cortex, often accompanied with collapse of some cells to form gas filled spaces. In order to cope up with the anaerobiosis in the root environment, many plants develop adventitious roots either from the submerged part or from just above the surface of water part of the stem or hypocotyl. Acceleration of adventitious root formation has been reported in most of the cereal crops such as maize, rice, barley, etc. In many tree species, adventitous roots develop from the hypertrophied lenticels. When adventitious roots are developed most plants partially recover from the inhibition of stem extension.

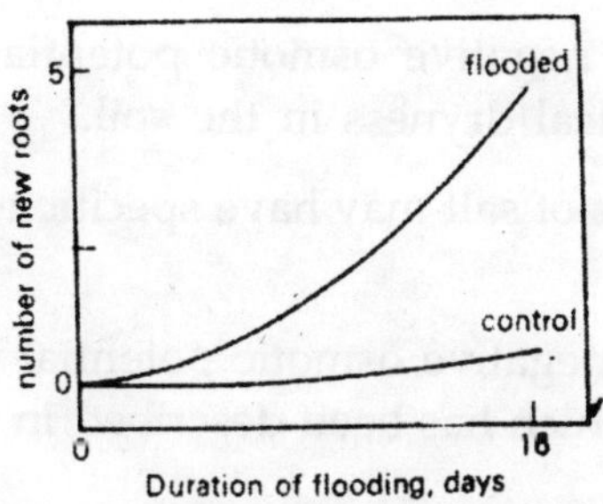

Effect of flodding on adventitious root formation in sunflower.

Water logging reduces leaf growth also. Leaf expansion can slow within 20 to 40 minutes of water logging.

Water logging induces epinasty in tomato. This has been shown to be due to increased ethylene production, during anaerobiosis. Anaerobiosis created by flushing of nitrogen also increases ethylene production in many tissues. Leaf abscission, reported in some species, is also due to increased ethylene production during water logging and anaerobiosis.

Soil Salinity: The condition of soil salinity is quite prevalent in many soils. It is created by the salts of sodium, primarily by NaCl, and also by Na_2CO_3, $NaHCO_3$ and Na_2SO_4. Salts of K, Ca and Mg are also involved in creating a salt stress on plants. Excess of salt in the soil has two types of effects on plants:

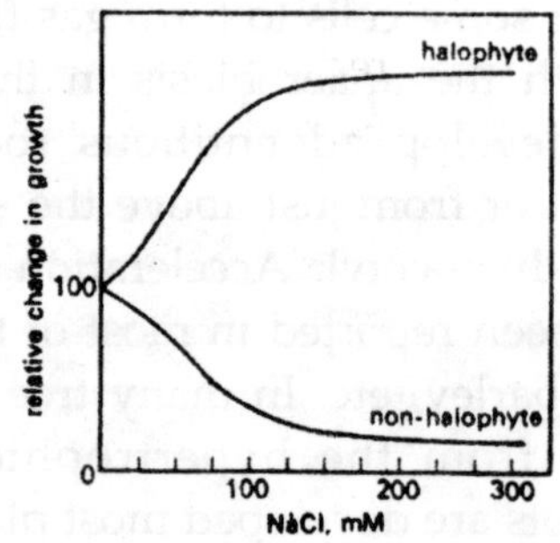

A generalized view of the effect of salt concentration on the growth of a halophyte and a non-halophyte.

(1) It creates negative osmotic potential and a kind of physiological dryness in the soil.

(2) The excess of salt may have specific toxic effect on the plant.

The effect of negative osmotic potential is similar to that of water stress, which has been described in earlier section of this chapter.

As far as the toxic effects of high concentrations of salts is concerned, plants differ according to the species, cultivar and even ecotypes. The growth of most non-halophytes, is decreased if salt concentration increases beyond a few millimolar, say about 10-20 mM. But again there is variation among species. Species such as onions, pea, groundnut, soybean, beans, etc. are very sensitive while lupines and *Festuca rubra* are comparatively resistant to salt Stress. For example, 150 mM NaCl solution may reduce the growth of some varieties of rice by as much as 80% within a few weeks, while that *of Festuca rubra* is inhibited by only about 20%, in similar conditions. In some species, gait hardiness can be induced by exposure to saline conditions.

Halophytes can tolerate and grow upto about 20% of salt concentration, although most grow in an environment containing 2 to 6% salt. Their growth increases with increase in salt concentration upto a limited range, although the growth of non-halophyte is severely reduced under similar conditions. These halophytes can accumulate high concentration of salt in their tissues. Osmotic potential of some halophytes is as low as -170 bars as compared to -20 to -30 bars in most non-halophytes.

As described earlier, some non-halophytes also have some degree of salt tolerance. Depending upon the mechanism of salt tolerance, there are two types of salt tolerant plants: (i) Salt excluders and (ii) Salt includers.

Salt excluders, transport very small amount of salt to the shoots. They either do not absorb it from the environment or accumulate in roots. Most salt tolerant, legumes and other cereals are salt excluders.

In salt includers, high amounts of salts are absorbed and stored in stem and leaves. Halophytes are in this category. Many salt includers are succulent, probably because of accumulation of salt in large vacuoles of the mesophyll cells. Some salt including species have special glands on the leaf surface that excrete salt at a high concentration.

Chemical Growth Regulators

Over the past 50 years or so, it has become increasingly obvious that the growth and development of plants are controlled by very low concentrations of a number of chemical agents. These chemicals usually have been introduced in agriculture as fungicides, pesticides, insecticides, etc. Some of these behave as endogenous growth hormones. A few of these chemicals and their effects on growth are described in the following paragraphs.

Phenolics: Phenolics are endogenous substances produced in secondary metabolic reactions. They are localised in chloroplast envelope, vacuole or cell wall. These compounds have a aromatic ring with —OH group (phenolic group) and depending upon the number of —OH group, they may be monophenols, diphenols and polyphenols.

Most of the phenolic compounds such as caffeic acid, p-coumaric acid, hydroxybenzoic acid, etc. inhibit growth. But many at lower concentration (say less than 0.5 mM) increase plant growth, as has been observed in maize with salicylic acid. This stimulation of growth has been shown to be partly due to the increased assimilation of nitrate by the maize plants.

Morphactins: Chemically, these are either flurenol or chloroflurenols. The chloro acid is the most active form of morphactins; Morphactins have diversified effects on plants and are known to interact with endogenous plant hormones. They are known to promote extension growth of internodes, opening of hypocotyl hook in seedlings, etc. Growth of wheat coleoptile segments is also promoted by morphactins. However, in many other studies morphactins have been found to be inhibitory to growth.

2-chloro 9-hydroxyfluorene 9-carboxylic acid (Chloroflurenol)
9-hydroxyfluorene 9-carboxylic acid (Flurenol)

Tricontanol: It is a high m.w. alcohol. It promotes growth in many herbaceous plants. The promotion is associated with increased CO_2 assimilation, water uptake and protein content.

Mixatalol: It is a mixture of long chain aliphatic alcohols varying in chain length from C-24 to C-32. It is known to increase growth of the mustard seedlings.

DCPTA: Chemically, it is amine: 2-(3,4-dichlorophenoxy)-triethylamine. It increases growth and yield of a wide variety of plants.

Monoethanolamine: Foliar spray of this chemical increases growth and yield and promotes the growth of the basal part of stem in rye and wheat. Growth retardants. There are many agro-chemicals which basically inhibit the plant growth.

However, at low concentrations some of these even increase the growth. Some of the commonly used growth retardants and their chemical names are as follows:

Ancymidol	:	α-cyclopropyl-α-(4-metnoxyphenyl) 5-pyrimidine methanol
AMO 1618	:	Ammonium (5-hydroxycaravacryl) trimethyl chloride carboxylate
CCC	:	Also known as chlormequat or chlormequat chloride. It is 2-chloroethyl trimethyl ammonium chloride.
CEPA	:	2-chloroethyl-phosphonic acid
Diaminozide	:	Succinic acid-2,2-dimethyl hydrazide (SADH)
EHPP	:	Ethyl hydrogen 1-propyl phosphonate
FTuoridamid	:	3-trefluomethyl-sulfonamide-p-acetoluidide
Paclobutrazol	:	(2RS,3RS)-l-(4-Chlorophenyl)-4,4-dimethyl-2- (1,2,4-triazol-l-yl-) pentan-3-ol
PPA	:	1-propyl phosphoric acid
Phosphon-D	:	2,4-dichlorobenzyle tributyl phosphonium chloride
RH 531	:	Sodium-l-(p-chlorophenyl)-l,2-dihydro-4,4-dimethyl 2-oxonicotinate
TIBA	:	2,3,5 tri-iodo benzoic acid
Triapenthenol	:	l-cyclohexyl-4,4-dimethyl-2-(1,2,4-triazol-l-yl) l-penten-3-ol

Triazole Derivatives: Some of the triazoles with plant growth activity are: Dichlorobutrazol, Etaconazole, Flusilazole, Triadimefon, Triadimenol, etc. They also protect the plants against stress injuries, such as water stress, air pollution stress and pathogens.

Most of the growth retardants are used in agriculture to regulate crop growth.

The major effect of retardants is to shorten internode length and plant height and generally to reduce lodging, specially in cereal grain crops and flax.

Some of these also cause darker green colouration by increasing chlorophyll content. Leaf area, light interception and product yield are usually not reduced by treatment.

Position of the Selected Families: In the following table the position of the selected families treated in this book is given in five major systems of angiosperm classification, *viz.* by Bentham and Hooker, Engler and Prantl, Hutchinson, Cronquist, and Takhtajan.

The system of Bentham and Hooker is that presented in *Genera Plantarum* (1862-1883).

In the Polypetalae and Gamopetalae cohorts of Bentham and Hooker are considered as equivalent to orders and in the Monochlamydeae and Monocotyledones their series are treated as orders in modern sense.

The 'orders' of Bentham and Hooker correspond to families in modern sense.

The system of Engler and Prantl is that presented in the 12th edition of *Syllabus der Pflanzenfamilien* edited by H. Melchior (1964).

The system of Hutchinson is that presented in *The Evolution and Phylogeny of Flowering Plants* (1969).

The system of Cronquist is that presented in *An Integratea System of Classification of Flowering Plants* (1981).

The system of Takhtajan is that presented in *Botanical Review* (1980).

	Bentham and Hooker	Engler and Prantl	Hutchinson	Cronquist	Takhtajan
DICOTYLEDONS					
Order: Ranales					
1.	Ranunculaceae	Ranunculales	Ranales	Ranunculales	Ranunculales
2.	Magnoliaceae	Magnoliales	Magnoliales	Magnoliales	Magnoliales
3.	Annonaceae	Magnoliales	Annonales	Magnoliales	Magnoliales
4.	Nymphaeaceae	Ranunculales	Ranales	Nymphaeales	Nymphaeales
Order: Parietales					
5.	Papaveraceae	Papaverales	Rhoeadales	Papaverales	Papaverales
6.	Fumariaceae (*in* Papaveraceae)	*in* Papaveraceae	Rhoeadales	Papaverales	Papaverales
7.	Brassicaceae (Cruciferae)	Papaverales	Cruciales	Capparales	Capparales
8.	Capparidaceae	Papaverales	Capparidales	Capparales	Capparales
9.	Violaceae	Violales	Violales	Violales	Violales
Order: Caryophyllinae					
10.	Caryophyllaceae	Centrospermae	Caryophyllales	Caryophyllales	Caryophyllales
11.	Portulacaceae	Centrospermae	Caryophyllales	Caryophyllales	Caryophyllales
Order: Malvales					
12.	Malvaceae	Malvales	Malvales	Malvales	Malvales
13.	Sterculiaceae	Malvales	Tiliales	Malvales	Malvales
14.	Tiliaceae	Malvales	Tiliales	Malvales	Malvales
Order: Geraniales					
15.	Oxalidaceae (in Geraniaceae)	Geraniales	Geraniales	Geraniales	Geraniales

Contd...

	Bentham and Hooker	Engler and Prantl	Hutchinson	Cronquist	Takhtajan
16.	Geraniaceae	Geraniales	Geraniales	Geraniales	Geraniales
17.	Rutaceae	Rutales	Rutales	Sapindales	Rutales
18.	Meliaceae	Rutales	Meliales	Sapindales	Rutales
Order:	Celastrales				
19.	Vitaceae (Ampelideae)	Rhamnales	Rhamnales	Rhamnales	Rhamnales
Order:	Sapindales				
20.	Sapindaceae	Sapindales	Sapindales	Sapindales	Sapindales
21.	Anacardiaceae	Sapindales	Sapindales	Sapindales	Rutales
Order:	Rosales				
22.	Fabaceae (Leguminosae)	Rosales	Leguminales	Rosales	Fabales
23.	Rosaceae	Rosales	Rosales	Rosales	Rosales
Order:	Myrtales				
24.	Combretaceae	Myrtiflorae	Myrtales	Myrtales	Myrtales
25.	Myrtaceae	Myrtiflorae	Myrtales	Myrtales	Myrtales
26.	Lythraceae	Myrtiflorae	Lythrales	Myrtales	Myrtales
Order:	Passiflorales				
27.	Cucurbitaceae	Cucurbitales	Cucurbitales	Violales	Cucurbitales
Order:	Umbellales				
28.	Apiaceae (Umbelliferae)	Umbelliflorae	Umbellales	Umbellales (Araliales)	Cornales

Contd...

	Bentham and Hooker	*Engler and Prantl*	*Hutchinson*	*Cronquist*	*Takhtajan*
Order:	Ficoidales				
29.	Cactaceae	Opuntiales	Cactales	Caryophyllales	Caryophyllales
Order:	Rubiales				
30.	Rubiaceae	Gentianales	Rubiales	Rubiales	Gentianales
Order:	Asterales				
31.	Asteraceae (Compositae)	Campanulales	Asterales	Asterales	Asterales
Order:	Ebenales				
32.	Sapotaceae	Ebenales	Ebenales	Ebenales	Ebenales
Order:	Gentianales				
33.	Apocynaceae	Gentianales	Apocynales	Gentianales	Gentianales
34.	Asclepiadaceae	Gentianales	Apocynales	Gentianales	Gentianales
Order:	Polemoniales				
35.	Boraginaceae	Tubiflorae	Boraginales	Lamiales	Polemoniales
36.	Convolvulaceae	Tubiflorae	Solanales	Polemoniales	Polemoniales
37.	Solanaceae	Tubiflorae	Solanales	Polemoniales	Scrophulariales
Order:	Personales				
38.	Scrophulariaceae	Tubiflorae	Personales	Scrophulariales	Scrophulariales
39.	Bignoniaceae	Tubiflorae	Personales	Scrophulariales	Scrophulariales
40.	Pedaliaceae	Tubiflorae	Bignoniales	Scrophulariales	Scrophulariales
41.	Acanthaceae	Tubiflorae	Personales	Scrophulariales	Scrophulariales

Contd...

	Bentham and Hooker	*Engler and Prantl*	*Hutchinson*	*Cronquist*	*Takhtajan*
Order:	Lamiales				
42.	Verbenaceae	Tubiflorae	.Verbenales	Lamiales	Lamiales
43.	Lamiaceae (Labiatae)	Tubiflorae	Lamiales	Lamiales	Lamiales
Order:	Curvembryae				
44.	Amarantaceae	Centrospermae	Chenopodiales	Caryophyllales	Caryophyllales
45.	Chenopodiaceae	Centrospermae	Chenopodiales	Caryophyllales	Caryophyllales
46.	Polygonaceae	Polygonales	Polygonales	Polygonales	Polygonales
Series:	Achlamydosporeae				
47.	Loranthaceae	Santalales	Santalales	Santalales	Santalales
Series:	Unisexuales				
48.	Euphorbiaceae	Geraniales	Euphorbiales	Euphorbiales	Eupborbiales
49.	Moraceae (In Urticaceae)	Urticales	Urticales	Urticales	Urticales
50.	Casuarinaceae	Casuarinales	Casuarinales	Casuarinales	Casuarinales
MONOCOTYLEDONS					
Series:	Microspermae				
51.	Orchidaceae	Microspermae	Orchidales	Orchidales	Orchidales
Series:	Epigynae				
52.	Musaceae (*in* Scitamineae)	Scitamineae	Zingiberales	Zingiberales	Zingiberales
53.	Zingiberaceae (*in* Scitamineae)	Scitamineae	Zingiberales	Zingiberales	Zingiberales
54.	Amaryllidaceae	Liliflorae	Amaryllidales	Liliales (in Liliaceae)	Liliales

Contd...

	Bentham and Hooker	*Engler and Prantl*	*Hutchinson*	*Cronquist*	*Takhtajan*
Series: Coronarieae					
55.	Liliaceae	Liliflorae	Liliales	Liliales	Liliales
56.	Commelinaceae	Commelinales	Commelinales	Commelinales	Commelinales
Series: Calycinae					
57	Palmae (Arecaceae)	Principes	Palmales	Arecales	Arecales
Series: Nudiflorae					
58.	Araceae (Aroideae)	Spathiflorae	Arales	Arales	Arales
Series: Apocarpae					
59.	Alismaceae	Helobiae	Alismatales	Alismatales	Alismales
Series: Glumaceae					
60.	Cyperaceae	Cyperales	Cyperales	Cyperales	Cyperales
61.	Poaceae (Gramineae)	Graminales	Graminales	Cyperales	Poales

Other Aspects

Dicotyledones

Division I. Polypetalae

Flowers with both calyx and corolla and the petals are distinct.

Series I. Thalamiflorae

Flowers mostly regular and bisexual; stamens hypogynous and ovary superior.

Order 1. Ranales

Stamens usually numerous; carpels free and embryo minute in fleshy albumen.

Ranunculaceae: Herbs or climbing shrubs; leaves radical or alternate; flowers regular or irregular; 1 or 2-sexual; sepals 5 or more usually deciduous, often petaloid; petals absent or 3-5 or more, imbricate; stamens indefinite; carpels usually numerous and free; fruits of numerous 1-seeded achenes or many seeded follicles, rarely a berry.

Magnoliaceae: Trees or shrubs with alternate, simple, exstipulate leaves; flowers spiral or spirocyclic; floral axis convex or elongated; perianth usually undifferentiated; stamens and carpels numerous and free; fruit follicular or samaroid or fleshy syncarp.

Annonaceae: Trees or shrubs with alternate, simple and exstipulate leaves; flowers usually trimerous; stamens and carpels usually many; fruits of one or many seeded indehiscent carpels.

Nymphaeaceae: Aquatic herbs with usually peltate leaves; sepals 3-5; petals 3-5 or many; stamens many; carpels 3 or more in one whorl, free or connate or irregularly sunk in pits on thalamus; fruit a spongy berry or of nuts.

Order 2. Parietales

Stamens definite or numerous; carpels united into a unilocular ovary with parietal placentation.

Papaveraceae: Herbs with milky or coloured juice and alternate exstipulate leaves; flowers actinomorphic; sepals usually 3 and caducous; petals twice as many as the sepals; stamens indefinite; ovary unilocular with many ovules on 2-5 parietal placentae; fruit a capsule.

Fumariaceae: Herbs with alternate leaves; flowers zygomorphic; sepals 2 and petals 4; stamens 6 in two phalanges with 1-2 nectaries at base; carpels 2, ovary unilocular with 1-α ovules on two parietal placentae; fruit usually 1-seeded nut.

Brassicaceae: Herbs with alternate exstipulate leaves; sepals and petals 4 each; stamens 6, tetradynamous; ovary usually 2-celled by the formation of a false septum; fruit usuallv a 2-valved pod.

Capparidaceae: Trees, shrubs or herbs with alternate, simple or compound leaves; sepals and petals 4 each; stamens 4-many; ovules many on 2-5 parietal placentae; androphore and gynophore usually present; fruit capsular or a berry.

Violaceae: Herbs or shrubs with alternate and stipulate leaves; flowers actinomorphic or zygomorphic; stamens 5; ovary unilocular with 3 parietal placentae; fruit a 3-valved capsule.

Order 3. Polygalinae

Sepals and petals usually 5 each, stamens as many or twice as many as the petals; ovary usually bilocular.

Order 4. Caryophyllinae

Flowers actinomorphic; stamens usually twice as many as the petals; ovary unilocular with free-central placentation.

Caryophyllaceae: Herbs with opposite and simple leaves; sepals 4-5; petals 4-5; stamens usually twice the petals and obdiplostemonous; ovary unilocular with free-central placentation; capsule 2-6-valved.

Portulacaceae: Herbs with alternate or opposite leaves; sepals 2, connate below; petals 4-6; stamens 4-many; ovary half inferior; capsule dehiscing transversely.

Order 5. Guttiferales

Stamens usually numerous; ovary 3-many locular; placentation axile.

Order 6. Malvales

Stamens numerous, monadelphous; ovary 3-many locular, placentation axile.

Malvaceae: Trees, shrubs or herbs with alternate, simple or rarely compound, stipulate leaves; sepals 5, valvate, free or connate; petals 5, twisted; stamens many, monadelphous and the stamen tube is adnate to the base of the petals; anthers one-celled; ovary 2-many celled, ovules 1 or more in each cell, placentation axile; fruit of distinct carpels or capsular.

Sterculiaceae: Herbs, shurbs or trees with alternate, simple or compound, stipulate leaves; flowers bi-or unisexual; sepals 5, more or less connate, valvate; petals 5, connate below or absent; stamens 5-20, more or less connate; anthers 2-celled; carpels 2-5; fruit of follicles or a loculicidal capsule.

Tiliaceae: Herbs, shrubs or trees with alternate, stipulate leaves; flowers mostly bisexual; sepals 5 or 4, free or connate; petals 5 or 4, distinct; stamens usually many, more or less distinct; anthers two-celled; ovary 2-5 celled; ovules 2-many in each cell; fruit a capsule or drupe.

Series II. Disciflorae

A conspicuous disc is present at the base of the ovary; ovary superior.

Order 7. Geraniales

Carpels several, syncarpous; ovules 1 or 2 in each locule, ascending or pendulous; raphae ventral.

Geraniaceae: Mostly herbs with simple and stipulate leaves; flowers regular or zygomorphic; sepals 5, free, equal or unequal; petals 5; stamens as many or two to three times as many as petals; ovary 5-locular with 1 or 2 ovules in each locule on axile placentation; fruit a schizocarp.

Oxalidaceae: Mostly herbs with compound, exstipulate leaves; flowers regular; sepals 5, imbricate, persistent; petals 5, imbricate or contorted, free or slightly united; stamens 10, obdiplostemonous, united below; carpels 5 with free styles; placentation axile; fruit a capsule.

Rutaceae: Trees or shrubs with usually compound, exstipulate, pellucid-punctate leaves; flowers bi-or unisexual; sepals 4-5, imbricate; petals 4-5, imbricate or valvate; stamens usually definite; carpels 4-5, free or connate; ovules 1-many in each cell; fruit usually a berry.

Meliaceae: Trees with alternate, compound, exstipulate leaves; stamens monadelphous; flowers uni-or bisexual; calyx 3-6-lobed; petals 3-6, imbricate; ovary 2-5-celled, placentation axile; fruit a capsule or drupe.

Order 8. Olacales

Ovules pendulous, raphae dorsal.

Order 9. Celastrales

Ovules erect, raphae ventral.

Vitaceae: Usually climbing shrubs or herbs with simple or compound, stipulate leaves; calyx cup-shaped; petals 4-5, distinct or connate, valvate; stamens as many and opposite the petals; ovary 2-6-locular with 1-2 ovules in each locule; fruit a berry.

Order 10. Sapindales

Flowers often irregular and unisexual; ovules usually ascending with a ventral raphae.

Sapindaceae: Trees, shrubs or herbs with simple or compound, exstipulate leaves; flowers often irregular; sepals 4-5, distinct and imbricate or connate and valvate; petals 4 5, distinct or sometimes wanting; disc annular or unilateral; stamens 5-10, distinct; ovary 1-6-locular with 1 ovule in each locule; fruit capsular or indehiscent

Anacardiaceae: Trees or shrubs with simple or compound, exstipulate leaves; flower uni-or bisexual: calyx 3-5 partite;

petals 3-5, free or rarely absent; disc annular; stamens 10 or 5 or less; ovary of 5 distinct one ovuled carpels or carpels united into a 2-5-celled ovary; fruit a drupe with 1 celled, 1 seeded stone.

Series III. Calyciflorae

Flowers perigynous or epigynous; disc rarely present.

Order 11. Rosales

Flowers bisexual, regular or zygomorphic; gynoecium of 1 or more carpels; styles usually distinct.

Fabaceae: Trees, shrubs or herbs, often climbing, with usually compound stipulate leaves; flowers regular or zygomorphic; sepals usually connate; petals 5, usually very unequal; stamens usually 10, diadelphous or monadelphous or distinct or many; gynoecium monocarpellary; ovary unilocular with marginal placentation; fruit usually a pod.

Rosaceae: Herbs, shrubs or trees with simple or compound stipulate leaves; flowers regular and bisexual; sepals and petals 5 each; stamens many, distinct; gynoecium of 1 or more free or connate carpels; ovules 1-2 in each carpel; fruit various.

Order 12. Myrtales

Flowers regular; ovary usually inferior; ovules 2 or more in each locule; style undivided.

Combretaceae: Trees or shrubs with simple exstipulate leaves; flowers regular and bisexual; calyx tubular, 4-5 lobed; petals 4-5; stamens twice as many as the petals; ovary unilocular with 2-5 pendulous ovules; fruit indehiscent, one seeded.

Myrtaceae: Trees or shrubs with simple usually gland-dotfed leaves; calyx 4-5-lobed; petals 4-5, free or united in a cup; stamens many; ovary 2-4-locular with many ovules in each locule; fruit indehiscent, 1-many seeded.

Lythraceae: Trees, shrubs or herbs with opposite exstipulate leaves; calyx 3-6-lobed, persistent; petals as many as calyx lobes, crumpled; stamens definite or many, inserted on the calyx tube; ovary 1-6-celled; fruit a many-deeded capsule.

Order 13. Passiflorales

Flowers bisexual or unisexual; ovary usually inferior, unilocular with parietal placentation, sometimes 3-locular by fusion of placentae; styles free or connate below.

Cucurbitaceae: Herbs climbing with the help of tendrils; flowers unisexual; calyx 5-lobed; petals 5, usually united; stamens 3 or rarely 5, on the calyx tube; anthers free or connate, lobes usually conduplicate; ovary unilocular with 3 parietal placentae meeting in the centre; fruit a pepo.

Order 14. Ficoidales

Flowers regular or subregular; ovary inferior to superior, unilocular with parietal placentation or 2-8-locular with axile placentation.

Cactaceae: Xerophytic shrubs or rarely trees; stem fleshy, of various shapes, rarely bearing green leaves; flowers usually solitary, sessile, regular or zygomorphic; perianth many, spirally arranged, often fused at the base to form a hypanthium; stamens numerous, epipetalous; gynoecium 2-many carpellary with 2-many parietal placentae bearing numerous ovules; fruit usually a berry.

Order 15. Umbellales

Flowers regular; ovary inferior, 1-many-locular with solitary pendulous ovule in each locule.

Apiaceae: Herbs with alternate simple or compound, exstipulate leaves; flowers in umbels, bisexual or polygamous; calyx 5-fid; petals 5, often inflexed; stamens 5; ovary bilocular with a single pendulous ovule in each locule; a prominent epigynous disc is present; styles 2; fruit schizocarpic with 2 one-seeded mericarps.

Division II. Gamopetalae

Corolla of united petals.

Series I. Inferae

Ovary inferior.

Order 1. Rubiales.

Flowers regular or zygomorphic; stamens epipetalous; ovary *o*-many locular with 1-many ovules in each locule.

Rubiaceae: Herbs, shrubs or trees with usually opposite and stipulate leaves; flowers regular: calyx limb 4-5-cleft; corolla tubular or rotate; stamens as many as the corolla lobes; ovary 2-many locular with axile placentation; fruit a berry, drupe or capsule, or of distinct cocci.

Order 2. Asterales

Flowers regular or zygomorphic; stamens epipetalous; ovary unilocular with a single ovule.

Asteraceae: Herbs, shrubs or rarely trees with alternate, exstipulate leaves; flowers regular or zygomorphic, unisexual or bisexual, arranged in a head and surrounded by an involucre of bracts; calyx usually modified into pappus; corolla of ray florets with the lobes connate in a strap, of disc florets with valvate spreading lobes; anthers connate; ovary unilocular with a solitary basal ovule; fruit an achene.

Order 3. Campanales

Flowers usually zygomorphic; ovary 2-6-locular with many ovules in each locule.

Series II. Heteromerae

Ovary usually superior; stamens as many or twice as many as the corolla segments; carpels more than two.

Order 4. Ericales

Flowers regular; stamens as many or twice as many as the petals; ovary 1-many locular with 1-many ovules in each locule.

Order 5. Primulales

Flowers regular; stamens usually equal to and opposite the corolla lobes; ovary unilocular with free central or basal placentation.

Order 6. Ebenales

Flowers regular; stamens usually more than corolla lobes; ovary 2-many locular with few ovules in each locule.

Sapotaceae: Trees with alternate, exstipulate leaves; calyx persistent; corolla lobes 5-many in 1 or 2 series; stamens 5, alternating with staminodes or many and without staminodes; ovary 6-8-locular, with 1 ovule in each locule; fruit a berry.

Series III. Bicarpellatae

Ovary superior; stamens as many, or fewer than corolla lobes; carpels usually 2.

Order 7. Gentianales

Flowers regular; stamens as many as the corolla lobes; leaves usually opposite.

Apocynaceae: Trees, shrubs or herbs with usually milky latex and opposite or whorled leaves; calyx 5-partite, often with scales or glands within; corolla salver-shaped or rotate, lobes contorted; ovary two, free, each unilocular with few to many ovules; fruit a berry or drupe or of two distinct or connate follicles.

Asclepiadaceae: Herbs or shrubs, usually twinning with opposite exstipulate leaves; calyx 5-lobed, lobes imbricate; corolla tube often with a ring of scales in the throat, lobes valvate or contorted; anthers adnate to the stigma forming a gynostegium; carpels 2 with free ovaries and styles; ovary unilocular with many ovules; fruit a pair of follicles.

Order 8. Polemoniales

Leaves generally alternate; flowers regular; stamens as many as the corolla lobes.

Boraginaceae: Herbs, shrubs or trees with opposite leaves; inflorescence corymbose, usually of scorpoid cymes; calyx usually 5-cleft, persistent; corolla tubular, funnel-shaped or rotate, lobes usually 5, imbricate; stamens usually 5, on the corolla tube; ovary bilocular with 2 ovules in each locule or tetralocular with 1 ovule in each locule; fruit a drupe or of 2-4 nutlets.

Convolvulaceae: Herbs or shrubs, often twinning; calyx persistent; corolla campanulate or funnel-shaped, lobes plaited or contorted in bud; stamens 5, on the corolla tube; ovary surrounded by an annular disc, bilocular with 2 ovules in each locule or 4 celled by a false septum with 1 ovule in each locule; fruit dry or fleshy, indehiscent or 2-4 valved or circumsciss, 1-4 seeded.

Solanaceae: Herbs or shrubs with usually alternate, simple and exstipulate leaves; calyx often persistent and enlarged, 5-cleft; corolla lobes 5, plaited or valvate in bud; stamens 5, on corolla tube; anthers sometimes opening by pores; ovary bilocular with many ovules in each locule; placentae swollen and septum oblique; fruit a berry or capsule.

Order 9. Personales

Flowers usually zygomorphic; corolla often bilipped; stamens generally fewer than the corolla lobes; ovules many in each locule.

Scrophulariaceae: Herbs with opposite or whorled exstipulate leaves; calyx 5-lobed or partite; corolla usually bilipped, lobes imbricate; stamens 4, didynamous, rarely 2 or 5; ovary bilocular; fruit a many seeded two-valved capsule.

Bignoniaceae: Trees with opposite compound leaves; calyx campanulate or spathulate; corolla campanulate or tubular, 5-lobed; stamens 4, didynamous or rarely 5; cushion-like or annular disc present; ovary bilocular with many ovules in each locule; fruit a capsule; seeds often winged.

Pedaliaceae: Herbs with opposite or alternate leaves; calyx 5-partite; corolla ventricose, sub-two-lipped, lobes 5, imbricate in bud; stamens 4, didynamous; ovary unilocular or 2-4-locular by confluence of parietal placentae; ovules 2 or more; fruit dehiscent.

Acanthaceae: Herbs or shrubs with opposite, simple, exstipulate leaves; calyx 4-5-partite; corolla 2-lipped, lobes imbricate or contorted; stamens 4 or 2 ovary bilocular with 1 or more ovules in each locule; fruit a loculicidal capsule.

Order 10. Lamiales

Corolla usually bilipped; stamens usually 4, didynamous or sometimes 2; ovary 2-4-locular with usually 1 ovule in each locule; fruit a drupe or nutlets.

Verbenaceae: Herbs, shrubs or trees with usually opposite, simple or compound leaves; calyx 2 or 4-6-partite; corolla regular or zygomorphic; stamens 4, didynamous; disc present; ovary 2 or 4 locular with 1-2 ovules in each locule; fruit usually a drupe.

Lamiaceae: Usually aromatic herbs or undershrubs; stem generally tetragonous; calyx 5-cleft, persistent; corolla usually bilipped; stamens 4, didynamous or rarely 2; disc present; ovary bilocular, becoming tetralocular by false septum, 1 ovule in each locule; style gynobasic; fruit of indehiscent nutlets.

Division III. Monochlamydeae

Flowers usually with one whorl of perianth, usually sepaloid.

Series I. Curvembryae

Terrestrial plants with bisexual flowers; stamens as many as perianth segments; ovule solitary; embryo curved.

Amarantaceae: Herbs or shrubs with opposite or alternate, exstipulate leaves; flowers uni-or bisexual; perianth members 5, scarious or herbaceous, imbricate, persistent; stamens 1-5, opposite perianth members; ovary unilocular with a solitary basal ovule; fruit usually an indehiscent utricle.

Chenopodiaceae: Herbs or undershrubs with alternate exstipulate leaves; flowers uni-or bisexual; perianth segments 3-5, sepaloid, distinct or connate, imbricate, persistent; stamens 5, opposite perianth lobes; ovary unilocular with a basal ovule; fruit a membranous utricle.

Polygonaceae: Herbs with alternate leaves and sheathing stipules; perianth usually 5-cleft, often coloured, imbricate, persistent; stamens 4-8; ovary unilocular; ovule basal; styles 2-3; fruit a trigonous or biconvex nutlet.

Series II. Multiovulatae Aquaticae

Aquatic plants with syncarpous ovary and numerous ovules.

Series III. Multiovulatae Terrestris

Terrestrial plants with syncarpous ovary and numerous ovules.

Series IV. Microembryae

Gynoecium syncarpous or apocarpous; ovule usually solitary; embryo very small.

Series V. Daphnales

Ovary superior, usually of 1 carpel; ovules solitary or few; perianth sepaloid.

Series VI. Achlamydosporeae

Ovary inferior, unilocular, 1-3-ovuled; seeds without testa and adnate to pericarp.

Loranthaceae: Branch parasites with opposite or alternate leaves; flowers uni-or bisexual; calyx truncate or absent; petals 3-5, distinct or connate; stamens as many and opposite the petals; ovary unilocular with a solitary erect ovule adnate to the ovary wall; fruit a 1-seeded berry.

Series VII. Unisexuales

Flowers unisexual; perianth sepaloid or much reduced or absent; ovary syncarpous or of 1-carpel; ovules 1-2 per carpel.

Euphorbiaceae: Herbs, shrubs or trees often with milky latex; perianth of a single whorl or of two distinct whorls or absent; stamens one or more; ovary usually trilocular with 1-2 pendulous ovules in each locule; fruit generally a regma.

Moraceae: Trees or shrubs often with milky latex and alternate stipulate leaves; flowers unisexual; perianth usually of 4 members, persistent; stamens 4, opposite to perianth members, bent inwards or straight in the bud; gynoecium of 2 carpels of which one shows various degrees of abortion; ovary unilocular and 1-ovuled; fruit an achene or drupe-like, but commonly a multiple fruit arises by union of fruits of different flowers.

Casuarinaceae: Trees or shrubs with xerophytic habit; plants monoecious; leaves scaly, borne in whorls at nodes; male flowers borne in catkin-like spikes; each has 2-leaved perianth, 2 bracteoles and a single stamen; female flowers in dense, spherical or ovoid heads; masked in the axil of a bract and has 2 bracteoles; carpels 2, syncarpous, posterior locule empty, anterior containing 2 ovules; fruit a 1-seeded nut.

Series VIII. Ordines Anomali

This series includes families of doubtful or unknown affinities.

Monocotyledones

Series I. Microspermae

Inner perianth petaloid; ovary inferior with 3 parietal or rarely axile placentae; seeds minute, exalbuminous.

Orchidaceae: Epiphytic or terrestrial herbs; flowers zygomorphic; perianth of two trimerous often petaloid whorls; posterior petal larger and is termed the labellum; stamen usually 1, confluent with the style in a column; pollen grains in waxy or powdery masses; ovary unilocular with 3 parietal placentae; fruit a loculicidal capsule.

Series II. Epigynae

Perianth partly petaloid; ovary inferior; endosperm abundant.

Musaceae: Gigantic herbs with pseudostems; flowers in spadix, unisexual or bisexual; perianth segments 5-6, 5 united and 1 free; stamens in two trimerous whorls, posterior often represented by a staminode; gynoecium tricarpellary, trilocular with 1-many ovules in each locule; fruit a berry.

Zingiberaceae: Perennial aromatic herbs often with tuberous roots; perianth in two trimerous whorls; stamens basically 6, the posterior member of the inner whorl is fertile and the other two are united to form the petaloid labellum, the anterior member of the outer whorl is always absent and the other two may be absent or present as large leafy staminodes; ovary 3 or 1 locular with numerous ovules; fruit a loculicidal capsule.

Amaryllidaceae: Perennial scapigerous herbs; perianth petaloid, 6-lobed; stamens 6, in two trimerous whorls; ovary trilocular with many ovules in each locule; fruit usually a capsule or sometimes indehiscent.

Series III. Coronarieae

Inner perianth petaloid; ovary superior; endosperm abundant.

Liliaceae: Herbs, rarely shrubs; perianth petaloid, 6-merous; stamens 6, in two trimerous whorls; ovary trilocular with 2 or more ovules in each locule; fruit a 3-celled berry or capsule.

Commelinaceae: Herbs or shrubs with alternate leaves having sheathing bases; flowers in cymes or panicles, more or less zygomorphic; perianth petaloid, 6-partite, in two series; stamens 6, all or only 3 perfect, filaments bearded; ovary trilocular and 1-few ovuled; fruit a loculicidal capsule.

Series IV. Calycinae

Perianth sepaloid, herbaceous or membranous; ovary superior, endosperm abundant.

Arecaceae: Shrubs or trees with alternate, variously compound leaves; flowers unisexual, enclosed in spathe; perianth 6-partite, in two series; stamens 6, in two whorls; ovary 1-3-locular or of 3 distinct or connate carpels with 1-2 ovules in each locule; fruit a berry or drupe.

Series V. Nudiflorae

Perianth none or very much reduced; gynoecium of one to several syncarpous carpels, ovary superior; ovules 1-numerous; endosperm usually present.

Araceae: Terrestrial or aquatic herbs or suffruticose climbers; flowers usually unisexual, on a spadix; perianth usually none; ovary 1-3-locular with 1 or more ovules in each locule; fruit 1-few seeded berry.

Series VI. Apocarpeae

Perianth in one or two whorls, or none; carpels free; ovary superior; seeds nonendospermic.

Alismaceae: Marsh or aquatic herbs; flowers uni-or bisexual; perianth 6-partite, in two series; stamens 3-many; ovary of few to many distinct carpels; ovules 1 or many; fruit of achenes or follicles.

Series VII. Glumaceae

Perianth of scales or none; ovary unilocular, 1-ovuled; seeds endospermic.

Cyperaceae: Grass-like herbs with terete or three-angled stems; flowers uni-or bisexual, in spikelets of imbricate bracts; perianth none or of scales or bristles; stamens 1-3, ovary superior, unilocular with a solitary erect ovule; stigmas 2-3, fruit indehiscent.

Poaceae: Mostly herbs; leaves distichous with sheathing bases; stems terete or compressed; flowers uni-or bisexual, in spikelets of imbricate bracts (glumes); perianth of 2-3 minute scales (lodicules) or none; stamens usually 3, versatile; ovary superior, unilocular with a single basal ovule; styles 2, feathery; fruit a caryopsis.

4

Floral Development

When a plant reaches a certain stage of vegetative growth, profound changes take place in its structure and function and it starts flowering. The apex of the branch on plant, is induced to form a flower instead of other branches or leaves. Significant changes take place at cellular and metabolic levels, during these modifications. All these changes are finally manifested into morphological changes; the vegetative bud is transformed into a reproductive or floral bud.

The flowering in plants is influenced by various environmental and nutritional factors. In most cases, a specific environment is necessary for flowering to occur. It should be noted however, that the specific environment acts as a floral stimulus, only when the plant has reached a certain stage of maturity. Very young plants, which are thought of as being in juvenile phase, can not flower, even if all other factors are favourable. They are receptive to floral stimuli when they have reached puberty.

Among various environmental factors, the day length, the quality and intensity of light and temperature are the most important variables affecting the timing and intensity of flowering. In fact, these factors are so intimately linked to the mechanism of flowering, that the study of their effects is essential for understanding the process of flowering.

Different Stages

Photoperiodism

Photoperiodism is defined as 'the response of plants to the timing of light and darkness'. Although the most important photoperiodic response in the plants is flowering, other responses such as pigmentation, branching, tuberization or dormancy can also be seen. The phenomenon of photoperiodism with respect to flowering is known since nineteenth century. In late nineteenth century, it was demonstrated that when an additional source of light such as incandescent filament lamp or incandescent gas light was used in the greenhouse, some vegetables flowered earlier than normal.

But the photoperiodism was named so by two plant physiologists, W.W. Garner and H.A. Allard (1920,22), working at the United States Department of Agriculture, Beltsville, Maryland (USA). They were studying growth and flowering behaviour in Biloxi soybean *(Glycine Max)* and in Maryland mammoth variety of tobacco *(Nicotiana tabaccum)*. Both these plants flowered only in a particular season, irrespective of their germination and growing season. They were short day plants and flowered during September-October, when the daily exposure to light was reduced below a certain critical duration. From several experiments Garner and Allard concluded that relative length of the day was the most important factor in growth and development of plants.

The phenomenon called photoperiodism was further characterized in many other investigations by other plant physiologists as veil. Now, it is known to occur in animals such as insects, birds, mammals is well. In these animals, reproductive development and behaviour is controlled by the day length.

Based on photoperiodic response of flowering, plants have been grouped into following four categories: (i) Short day plants, (ii) Long day plants, (iii) Indeterminate or day neutral plants and (iv) Intermediate plants.

Short Day Plants: Flower readily only when the photoperiod is shorter than a critical period. Under longer (than the critical period) photoperiods, these species or varieties do not flower and remain in the vegetative state. This critical day length or the photoperiod depends upon the species. For example, violet plants flower when the day length is shorter than 11 hours, while chrysanthemum flowers in day length shorter than 15 h. Both of these species, however, are short day plants, because they do not flower when the day length is more than the critical 11 or 15 h. Some examples of short day plants are given in table

Some Short Day Plants

Monocots	***Dicots***
Winter rice *(Oryza saliva)*	
	Bryophyllum *(Bryophyllum pinnatum)*
	Chrysanthemum *(Chrysanthemum* spp.)
	Cocklebur *(Canthium strumarium)*
	Cosmos *(Cosmos sulfureus)*
	Hemp *(Cannabis saliva)*
	Japanese morning glory *(Pharbitis nil)*
	Kalanchoe *(Kalanchoe blossfildiana)*
	Lespedeza *(Lespedeza stipulacea)*
	Morning glory *(Ipomaea purpurea)*
	Poinsettia *(Euphorbia pulcherrima)*
	*Strawberry *(Fragaria chiloensis)*
	*Tobacco *(Nicoliana tabaccum)*
	Violet *(Viola papilionacea)*

*Different varieties of these plants have different day length requirements.

Long day plants flower readily only under a range of photoperiod longer than the critical photoperiod, upto and including continuous illumination. Flowering is induced under very long photoperiods of continuous illumination. But a few long day plants may also flower slowly and less profusely under shorter photoperiods as well. Again the critical day length varies according to the species. For example, barley or wheat plant flower only when day length is more than 12 h, while spinach plants flower only when day length exceeds over 13 h. Some species such as *Agrosti palustris,* have a very long photoperiod, more than 16 h. Some long day plants are listed in table.

Some Long Day Plants

Monocots	***Dicots***
Barley *(Hordeum vulgare)*	Cabbage *(Brassica* spp.)
Bentgrass *(Agroslis palustris)*	Clover *(Jrifolium pratense)*
Oats *(Avena sativa)*	Cone flower *(Rudbeckia bicolor)*
Orchard grass *(Dactylis glomerata)*	Dill *(Anethum graveolens)*
Rye grass *(Lolium perenne)*	Henbane *(Hyoscyamus niger)*
Timothy *(JPhleum* spp.)	Hibiscus *(Hibiscus syriacus)*
Wheat grass *(Agropyron smithi)*	Petunia *(Petunia* sp.)
Wheat *(triticum vulgare)*	Radish *(Raphanus sativus)*
	Spinach *(Spinacea oleracea)*

Indeterminate or Day Neutral Plants: Flower readily over a wide range of day length from relatively short day length to continuous illumination (Table).

Some Day Neutral Plants.

Monocots	*Dicots*
Blue grass *(Poa annua)*	Azalea, coral bell *(Rhododendron* spp.)
Maize *(Zea mays)*	Balsam *(Impateins balsamina)*

Contd...

Monocots	*Dicots*
	Bean (*Phaseolus* spp.)
	Buckwheat (*Fagopyrum tataricum)*
	Cotton (*Gossypium hirsutum)*
	Cucumber (*Cucumis sativus)*
	Potato (*Solanum tuberosum)*
	Tomato (*Lycopersicum esculentum)*

Intermediate Plants: Flower only under day lengths within a certain range and fail to flower under either longer or shorter photoperiods. In this category, some varieties of sugar cane, *Eupatorium hyssopifolium* and climbing hempweed *(Mikania scandens)* are included.

Another term ambiphotoperiodism has been used by some plant physiologists for photoperiodic responses of plants like, *Media elegans*. Ambiphotoperiodic plants remain vegetative in intermediate day length and flower only on longer or shorter photoperiods. In this respect, they are counterparts of the intermediate day length species.

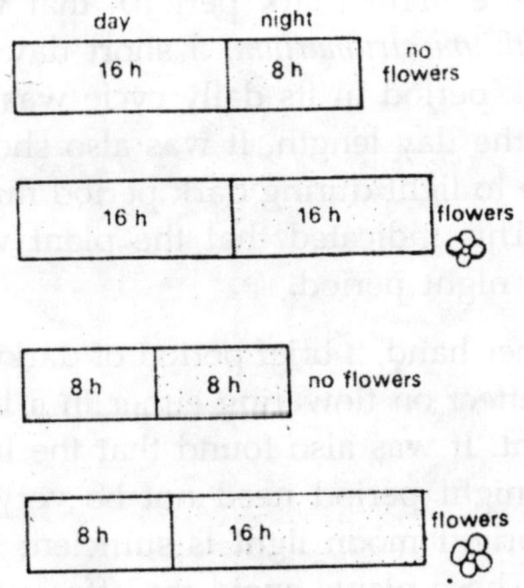

Effect of long and short night on flowering in Xanthium strumarium.

In all cases of photoperiodism studied under usual conditions of light-dark cycle totaling 24 h, there is a definite day length or photoperiod, in which a given species flowers. This is the critical day length of the species. Short day plants flower only when the day length (in a 24 h day/night cycle) is equal to or shorter than that critical photoperiod. For example, critical photoperiod for *Xanthium* (a short day plant) is about 15.5 h and it flowers only when day length is 15.5 h or less. Similarly long day plants flower only when exposed to day length more than a critical day length.

For example, *Lolium perenne* flowers only when the day length is 9 h or more and hence this plant is considered to be a long day plant. Thus, it is important to understand that the grouping of short or long day plants is on the basis of a definite or critical day length requirement and on the basis of whether the species flower in shorter or longer than the critical photoperiod.

Critical Night: To begin with, it was believed that in a photoperiodic response, it was the length of the day (light period) which determined the response. However, experiments by K.C. Hamner and J. Bonner demonstrated that in fact, it was he length of the night (dark period), that was critical. For example, *Xanthium strumarium*, a short day plant, flowered when the dark period in its daily cycle was more than 9 h, regardless of the day length. It was also shown that even a brief exposure to light during dark period nullified the effect of darkness. This indicated that the plant was sensing the uninterrupted night period.

On the other hand, a brief period of darkness during day time had no effect on flowering either in a long day or in a short day plant. It was also found that the light intensity to interrupt the night period need not be very high. In some plants, even bright moon light is sufficient to interrupt the night period. These plants avoid the effect of moon light by folding their leaves at night.

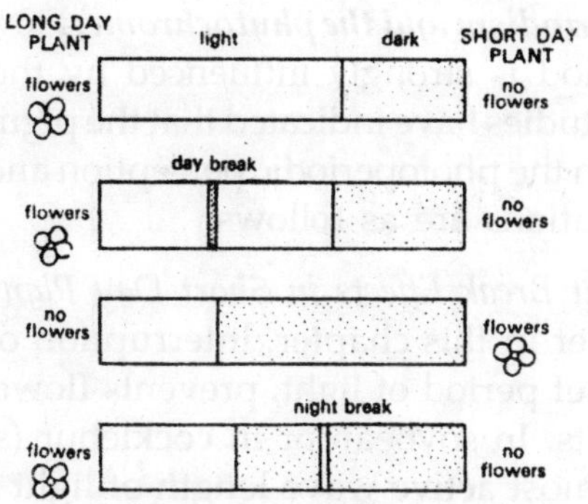

Effect of light and dark interruption on flowering in long and short day plants.

Because of the importance of dark period in flowering, some plant physiologists prefer to call short and long night plants instead of long and short day plants, respectively.

Site of Photoperiodic Perception: The perception of day or night length is confined to young expanded leaves. Even if one leaf of the plant is kept under the correct day length, the plant flowers, regardless of the conditions surrounding the rest of the plant. If all the leaves are removed from a plant, it becomes insensitive to the photoperiod. However, if even one leaf is left on the plant, and given an appropriate photoperiodic treatment, the plant flowers.

A single leaf may vary in its sensitivity to photoperiodic perception, during its development. The maximum sensitivity is reached when the leaves are about half their final size. Thereafter, the sensitivity declines, and the older leaves show low sensitivity. However, in some species, the maximum sensitivity may be attained much sooner, as in *Anagalis arvensis*, where leaves are most sensitive when only one twentieth of their final size. In *Lolium temulentum*, the photoperiodic maturity is retained by mature leaves.

There are also a few reports in the literature which indicate that buds with unexpanded leaves can also be sites of photoperiodic perception.

Photoperiodism and the phytochrome: Flowering in response to photoperiod is strongly influenced by the quality of light. Numerous studies have indicated that the pigment phytochrome is involved in the photoperiodic perception and flowering. Some such observations are as follows:

1. *Night Break Effects in Short Day Plants:* As mentioned earlier in this chapter, interruption of night period by a brief period of light, prevents flowering in short day plants. In soybean or in cocklebur (short day plants), the most active wave length of light in preventing the flowering, is the red region of the spectrum. Blue light with short wave lengths, is not effective. Also there is a rapid decline in the efficiency in far-red region. We have seen earlier that the phytochrome is converted into its active form in red light and into inactive form in far-red light. Thus, the effect of red, far-red light on night break effects on flowering in short day plants is consistent with the involvement of phytochrome.
2. *Promotion of Flowering in Long day Plants:* In many long day plants, including barley and *Hyoscyamus niger*, red light is the most effective in inducing flowering. The effect is reversible by far-red light. Further, photoperiodic effects are brought about by lights of very low intensities.

   Since green leaves were the sites of photoperiodic perception, it was also thought that photosynthesis might be involved in photoperiodism. However, this idea has been negated on following counts:

   (1) Photoperiodic responses can be evoked by lights of very low intensities, some as low as that of moonlight. Such lights are unable to initiate photosynthesis.

   (2) In photoperiodic responses, the more effective wave length of light is the red light, other regions of spectrum are ineffective. On the other hand, photosynthesis can take place in blue light as well, besides red light.

(3) The photoperiodic responses are red, far-red reversible, while it is not so in photosynthesis. In photoperiodic induction of flowering, the effect of red light can be completely nullified, if followed by a period of far-red irradiation. For example, a night break with red light inhibits flowering in short day plants, but the inhibitory effect is reversed, if far-red is given after red light exposure.

Time Measurements in Photoperiodism: As mentioned earlier, the critical day or critical night length depends upon the species. How precise this critical day or critical night period has to be? Many plants are able to sense a difference in day length of a few minutes, indicating that they are able to measure time with great accuracy. Two models have been proposed regarding the photoperiodic time sensing by the plants: (1) Hour glass timer and (2) Circadian clock.

1. *Hour Glass Timer:* This model was proposed following the study of flowering in short day plants such as *Xanthium, Pharbitis* and *Chenopodium.* In these species, dark period is more important than light period for flowering to occur. Flowering never occurs unless a certain minimum of dark period is applied. Not only this, in *Pharbitis,* the number of flowers produced on a given plant is proportional to the period in dark, beyond critical length.

   The change taking place during light-dark perception is visualised as an 'hour glass timer'. The process of change can be reversed as the hour glass timer can be inverted. Accordingly, it has been suggested that an unknown substance disintegrates over a period of time in dark until it decreases below a threshold level. At this time, other reactions may start which may eventually initiate flower formation. When the plant is exposed to light, this hour glass is inverted *i.e.,* the unknown substance starts accumulating. As described later in this chapter, the biochemical basis of this 'hour glass timer' is most likely the phytochrome interconversions; Pr ⇄ Pfr.

2. *Circadian Clock:* In the early 1930's, E. Bunning proposed a theory involving the same clock for circadian (circa + dien = approximately 24 h) rhythm and for the time measurement in photoperiodism. According to the theory, the metabolism of plants oscillates between two separate phases: the photophile (light loving) and skotophile (dark loving) phase. When the plant is in photophile phase, exposure to light increases its various processes including flowering. In skotophile phase, light inhibits flowering and other plant processes.

   The photoperiodism is executed by an endogenous free running oscillation between these two phases at a regular interval of approximately 12 hours. The photophile phase is believed to be starting with the onset of dawn (light) and skotophile phase starts after about 12 hours. If the light period is extended until skotophile phase, then flowering in short day plants is inhibited and that in long day plants is promoted. In fact short day plants may have more than 12 hours of skotophile phase. Similarly, long day plants may have a longer photophile phase. In long day plants, it is assumed that the beginning of photophile is delayed until about 12 hours after transfer in light.

Some of the experimental evidences favouring the circadian clock model are as follows:

(1) By varying the length of both light and the dark period, it has been established that the total duration of the cycle (light + dark) is an essential factor in determining the flowering responses in certain species. In Biloxi soybean for example, maximum flowering occurs under 24 hour cycle, provided the photoperiod is of 4 to 12 hours. The flowering is considerably reduced in a cycle of 34 hours.

(2) Night break experiments also suggest the existence of a circadian clock timer. In short day plants, *Kalanchoe,*

soybean and *Perilla,* with a 48 hour cycle initiated with a short light period of 6 to 10 hours, light interruptions of 30 minutes to 4 hours inhibit flower formation, when given near the beginning or end of the dark period. In long day plants, light interruptions near the beginning or end of the dark period promote flowering.

As far as the biochemical nature of the timer or clock is concerned, phytochrome seems to be the most important molecule, although the involvement of other molecules can not be ruled out. Borthwick and Hendricks (1960) were the first to propose that the disappearance of phytochrome far-red (Pfr) and reappearance of phytochrome red (Pr) constituted the hour glass timer. They suggested that the induction of process can not start before all the Pfr has been converted to Pr. This is further supported by the observation that far-red light given at the beginning of dark period reduces the critical night length in various short day plants.

The critical night length represents the lime taken for the concentration of Pfr in the leaves to fall below a threshold value which no longer inhibits flowering in short day plants nor promotes flowering in long day plants.

Biochemistry of Flowering: The plants which are sensitive to photoperiods, a definite chemical stimulus is produced in the leaves, when the plants are exposed to the desired photoperiod. The stimulus is then translocated to the shoot apex where it evokes flowering. Thus, as far as the biochemistry of flowering is concerned, it can be divided into two steps:

(1) The production of floral stimulus and

(2) Evocation of flowering.

Floral Stimulus

Grafting experiments suggest that a chemical is produced when plants are placed on a photoinductive cycle (a cycle which induces flowering). A plant kept in a photoinductive cycle flowers, because this chemical is present in that plant. One kept in a non-inductive cycle does not flower.

However, if a leaf from a photoinduced plant is removed and then grafted on a non-induced plant, then this plant also flowers. Apparently a chemical substance is produced during photoinductive cycle, which is transmitted during grafting of leaf from plants kept in such a cycle.

Several attempts have been made to extract and characterise the floral stimulus from flowering plants or photoinduced plants. However, attempts have met with limited success only. K.C. Hamner and J. Bonner (1938) tested some 246 different kinds of extracts from short-day, long day and day neutral plants, but none had any flower evoking effect in non-inductive plants.

Numerous known and unknown substances have also been tried by various investigators since then; but there is no uniformity in either response or chemical nature of the substance. Nevertheless, florigenic acid, gibberellins, sterols, phenolic acids and several other compounds have been shown to act as a flower stimulus in many species. Some of these are described in the following paragraphs.

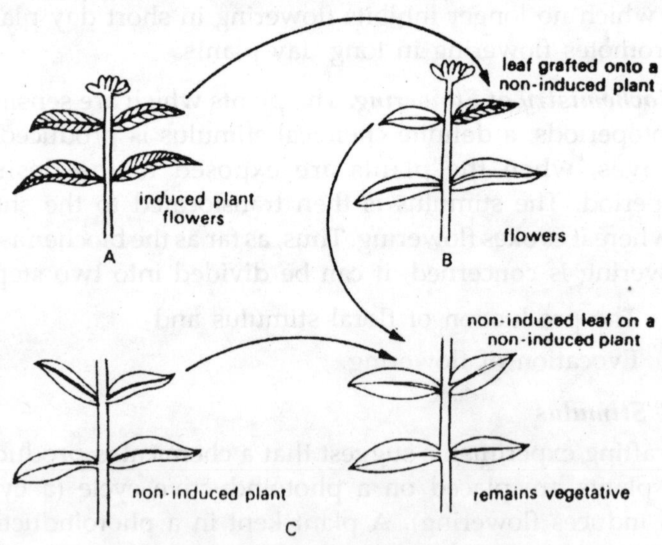

Induction of flowering by grafting a leaf from photo-induced to non-induced plant.

Florigen: The name 'florigen' (Greek meaning flower generating) was proposed by the Russian plant physiologist, M. Chailakhyan in 1936, for the unknown chemical stimulus which could act as a floral inducer. The most significant evidence for the existence of such a substance comes from interspecific grafting experiments. Flowering in a species can be brought about in a non-inductive cycle, if a leaf or twig from a photoinduced species is grafted on that.

Such results clearly indicate that floral stimulus produced by plants of one photoperiodic response can be transmitted by grafting to other plants, which may have different photoperiodic requirements. But florigen as proposed by Chailakhyan, has never been isolated and fully characterized and remains hypothetical. Either it does not exist in plants or it is too unstable to be isolated. Therefore, we now talk of 'florigen concept' instead of florigen. The concept relates to the existence of a chemical substance or substances involved in inducing flowering.

It would be rather useful to give historical perspective of the florigen concept, to understand its involvement in flowering. After Chailakhyan advanced the idea of existence of florigen, several plant physiologists tried to isolate and characterise the chemical. In 1951, R.G. Lincon, D.L. Mayfield and A. Cunningham obtained a chemical in crystalline form from flowering *Xanthium* and sunflower plants.

The substance did not exist in vegetative plants and it was found to be a water soluble organic acid. It induced flowering only when applied to the leaves and was ineffective when applied to the buds. It was thought to be a precursor of florigen or a substance necessary for florigen synthesis. In 1970, H.K. Hodson and K.C. Hamner tested the efficacy of florigen preparation in *Xanthium* as well as *Lemna.*

In experiments, the chemical induced flowering in about 75% of the *Xanthium* Plants, to which it was applied, and it induced flowering in about 50% of the *Lemna* plants. However, in *Xanthium,* the chemical was effective only when gibberellin was also added, although the gibberellins alone could not induce flowering. In *Lemna*, the substance induced flowering only when it was applied without gibberellins, added gibberellins made the extract totally ineffective.

The florigen was believed to be synthesized in the photoinduced leaves and from there moved on to the shoot apices for floral evocation, through phloem. S. Imamura and A. Takimoto (1955) studied the translocation rate of this stimulus. In Japanese morning glory, the rate was observed to be slow, 2.5 to 3.0 nm per hour as compared to more than 200 nm per hour for most sugars.

Thus, the florigen was not just a sugar synthesized in photoinduced leaves. In other studies also, exogenous application of sugars or amino acids did not induce flowering in non-induced plants. This strengthened the belief that the florigen was not just a general type of synthetic output of the photoinduced leaves.

In search for the chemical identity of the florigen, plant physiologists focused their attention to the effect of plant hormones on flowering. Several hormones have flower

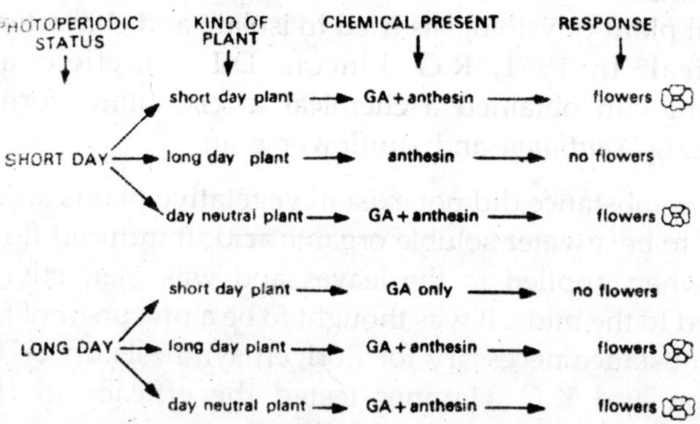

Diagrammatic representation of the formation and involvement of GA and anthesin in flowering.

promoting effects, although none can qualify as a florigen because of its limited effect. Gibberellins have the most spectacular effect, as they induce flowering in most long day plants, but not in short day plants. Following the discovery of effect of gibberellins on flowering, Chailakhyan modified his florigen concept. He proposed that florigen is two hormones rather than one, a gibberellin and a hypothetical hormone he called *anthesin.* He suggested that long day plants could produce anthesin under-any day length but gibberellins only under long days, that short day plants could produce gibberellins under any day length but anthesin only under short days, and that day neutral plants could produce both under any day length.

A plant could flower only when both gibberellin and anthesin were present. But again anthesin is just as hypothetical and elusive as florigen has been.

Gibberellins and the Florigen Concept: Exogenous application of gibberellins (GA) is known to induce flowering in many long day plants, under short day conditions. Natural GAs from induced long day plants also cause flowering in non-induced plants. In some plants increase in GA following photoinductive cycle has also been observed.

Although, GA can induce flowering, they can not be considered to be the primary floral hormone or florigen, mainly because of the following reasons;

(1) In long day plants, induction of flowering by long days and GA appear to be different. When these plants are kept in long days, differentiation of floral primordia occurs simultaneously with stem elongation. However, in GA induced flowering first stem elongation takes place and then floral primordia are formed. Thus, the effect of GA appears to be indirect; through stimulated growth and differentiation.

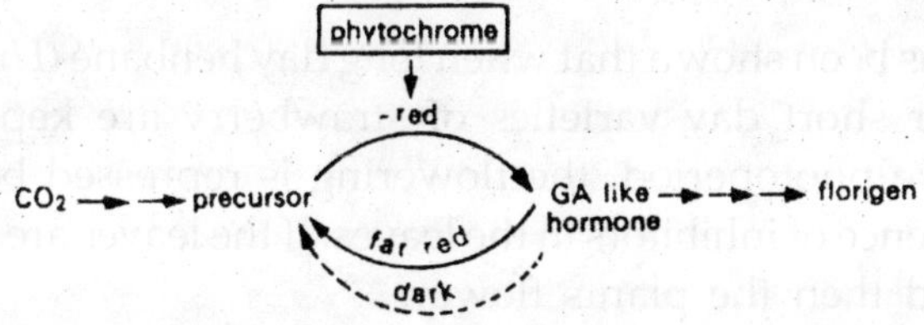

(2) Gibberellins have been unable to induce flowering in short day plants under non-inductive cycle.

Nevertheless, GAs seem to play an important role in flowering. It may be one of the several known and unknown compounds involved in flowering. It may be directly involved in the formation of florigen or any such floral stimulus.

A model based on the role of phytochrome, GA and florigen in the photoperiodic induction of flowering can be proposed. According to the model, photoperiodic reactions involving assimilation of CO_2 causes the build up of a precursor of GA or GA like hormone. This precursor may be converted to GA. Red light promotes this conversion and far-red or dark causes reconversion of the hormone back to the precursor. This reversible conversion is mediated via phytochrome.

The phytochrome may also be considered to be either an integral part or intimately linked to the precursor and hormone. This GA or GA like hormone then causes the formation of florigen either directly or indirectly. It is assumed that a high level of GA like hormone must be maintained in long day plants for the production of florigen. In short day plants, the situation is reverse, a low level of GA is required for florigen. However, once enough florigen is produced, flowering will occur in both long day and short day plants.

Flowering Inhibitor: Some experiments indicate that flowering in plants is controlled by the presence of a flowering inhibitor, either independently or in the presence of the floral stimulus. The inhibitor is apparently present in non-induced plants and either disappears or diminishes when the plant is kept on an inductive cycle.

It has been shown that when long day henbane (*Hyoscyamus niger*) or short day varieties of strawberry are kept in non-inductive photoperiod, the flowering is repressed because of the presence of inhibitors in the leaves. If the leaves are however, removed then the plants flower.

Existence of a specific transmissible inhibitor has been demonstrated in non-induced leaves of *Coleus, Fragaria, Lolium, Rottboellia* and tobacco. Y. Ogawa and R. W. King (1990) have also indicated the presence of inhibitor in the cotyledons of *Pharbitis nil* (Japanese morning glory).

It is a short day plant and does not flower when the plant is kept under continuous illumination. However, if the cotyledons of the seedling are darkened by covering with aluminium foil for 13 to 15 h, then plants flower.

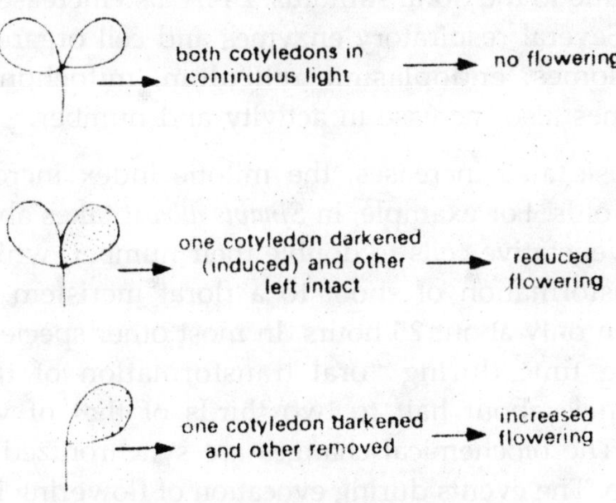

Increased flowering in* Pharbitis nit *by removal of the non-induced cotyledon

Ogawa and King (1990) have found that if out of two cotyledons only one is kept in dark (photoinduced) and the other cotyledon is left in light (non-induced), then the flowering is reduced as compared to when the other cotyledon (non-induced) is removed. Apparently, when the other cotyledon is under non-inductive (light) cycle, it produces some kind of inhibitor which inhibits flowering in this plant.

Only a few attempts have been made to extract and isolate a floral inhibitor. R.J. Pryce (1972) isolated gallic acid from the short day plant *Kalanchoe blossfeldiana* growing in long days, which acted as a floral inhibitor.

Flowering in Practice

Evocation of Flowering

The transfer of the floral stimulus (florigen) from leaf to the apical meristem, brings about several biochemical and cytological changes in the meristem, and it is transformed into a floral bud. The changes are almost identical in most species.

One of the early changes is increased RNA synthesis. This possibly is due to derepression or induction of a polycistronic operon due to the floral stimulus. DNA also increases in some plants. Several respiratory enzymes and cell organelles such as ribosomes, endoplasmic reticulum, mitochondria and dictysomes also increase in activity and number.

Mitosis also increases, the mitotic index increasing by several folds. For example, in *Sinapis alba* it takes about 157 h for the vegetative cells to double their number, while during the transformation of shoot to a floral meristem, the cells double in only about 25 hours. In most other species, the cell doubling time during "oral transformation of the apical meristem is about half to two thirds of that of vegetative tissues. The biochemical changes are synchronized with the cell cycle. The events during evocation of flowering in Sinapis alba are summarised in table.

Sequence of Events During Floral Evocation in Sinapis Alba

Hours after beginning of the inductive long day	*Events*
0	Most cells in G_1 phase
10	Increase in RNA synthesis
18	Increase in mitochondria
22	Increase in succinic ehydrogenase activity
26	First mitotic peak

Contd...

Hours after beginning of the inductive long day	*Events*
27-30	RNA synthesis at minimum, most cells in G_1
46	Most cells in G_2
62	Second mitotic peak, nucleolar volume maximum dictysome number maximum, flower buds begin to differentiate

The events of evocation are similar in plants which show the ability to flower at seedling stage (*e.g., Pharbitis nil)* and in those requiring a period of ageing before they flower (*e.g., Xanthium).*

Initiation and Development of Flower: Once the biochemical requirements for evocation of flowering are completed, and the meristem has reached the point of no return, it develops either into an inflorescence (a cluster of flowers) or a solitary flower. In most plants, the pattern of flower initiation and development is almost similar. As an example of flower initiation in *Capsicum annum* (green pepper), the first microscopically visible change in the shoot apex is the change in its shape.

The apex almost becomes flat from conical, apparently due to the inhibition of growth in the central portion of the meristem. Some protuberances develop from this meristem in a whorled manner. Floral parts (sepals, petals, etc.) are formed due to the development of these protuberances. The outermost whorl of the protuberances for the sepals and next to it form petals and so on.

Most plants produce bisexual or perfect flowers containing functional male (stamens) and female (pistils) parts. Other species contain starninate (male) and pistillate (female) flowers only on different individual plants. Photoperiod, temperature and some growth hormones markedly influence the sexuality in many plants. In maize and in some cucurbits, the ratio of male to female flowers can be increased by long days, but in spinach (a long day plant), the opposite effect is observed.

In cucumber, cool nights decrease and warm nights increase the production of female flowers. In several other plants however, warm temperatures favour staminate and cold temperatures pistillate development. In cucumber and squash, the ratio of male to female flowers is decreased by auxins. Ethylene also stimulates the formation of female flowers in cucumber. Gibberellins have an opposite effect; they increase the ratio of male to female flowers in the cucumber.

At first, the floral parts are tightly enclosed within the outermost part, the sepals, constituting a floral bud. Subsequently, expansion of the flower bud into an open flower, occurs. This stage in the flower development is called *anthesis*. Prior to anthesis, and when flower has reached its maturity, several morphological and physiological changes, take place. The peduncle ceases to grow, the rate of water uptake into the flower increases greatly and there is a large increase in the rate of respiration.

The respiration of opening flowers is so intense that the heat released sometimes causes a significant rise in temperature. Until the flowers open, there is also an accumulation of inorganic and organic solutes in various parts of the flowers. In some plants, more than 30% of the dry weight of petals may be sugars, at anthesis. These decrease rapidly after anthesis. The cause of the flower opening is usually due to the differential growth of the inner and outer sides of the sepals and petals.

Induction of Flowering

Besides treatment with an appropriate photoperiod, several other environmental as well as chemical treatments induce flowering. Sometimes these treatments are absolutely essential, but in other cases, they just hasten or increase the flower formation. A few important factors are described in the following paragraphs:

Increased Carbon—Nitrogen Ratio: G. Klebs (1913) and E. Kraus and H.R. Kraybill (1918) had proposed that C/N ratio of the plant was an important determinant in flowering. Increase in flowering due to increased C/N ratio was subsequently observed in many species including *Pharbitis nil* and *Lemna paucicostata.* Some short day species of *Lemna* can be induced to flower even under continuous light by culturing them on a nitrogen free medium. Day-length independent flowering is also induced by several chemicals, including ferricyanide which inhibit nitrogen assimilation by the plants. On the other hand, treatment with ammonium ions, nitrite, glutamine or a mixture of amino acids, inhibit day length independent flowering.

Likewise, exogenous supply of sucrose, which would increase C/N ratio in the plant, increases flowering in many species including *Anagalis arvensis, Pharbitis nil,* etc.

Applied Plant Hormones and Growth Regulators: Gibberellins are the most effective plant hormone in inducing flowering. It substitutes for long day requirements and also the cold treatment requirement in several species. In conifers also, plants belonging to the family cupressaceae, and taxodiaceae, respond to applied gibberellins. A mixture of GA_4 and GA_7 induces flowering in the members of pinaceae also.

Some examples of long day plants which can be induced by gibberellins under non-inductive (short day) photoperiods are: *Arabidopsis thaliana, Cichorium endivia, Crepis parviflora, Hyoscyamus niger, Lactuca sativa, Nicotiana sylvestris, Rudbeckia bicolor, Spinacea oleracea,* etc.

Examples of some cold requiring biennials which can be induced to flowering by GA are: *Beta vulgaris, Brassica oleracea, Daucus carota, Myosotis alpestris, Viola tricolor,* etc.

Gibberellins usually do not induce flowering in short day plants under non-inductive conditions, although they may cause stem elongation.

Auxins are also known to induce flowering in pine-apple and they have been used commercially for this purpose, ever since their this effect was discovered in 1942. They have also been found to be effective on long day plants, Wintex barley and *Hyoscyamus niger*. However, in most other species, either they have no effect or inhibit flowering. It is possible that the inhibitory action of auxin on flowering is mediated via ethylene production, which is stimulated by auxin treatment.

Exogenous application of cytokinins is also known to induce flowering in many species, under non-inductive photoperiods, including *Chrysanthemum, Lemna paucicostata, Perilla, Pharbitis nil, Wolffia*, etc. However, they are known to inhibit flowering in *Chenopodium*.

Abscisic acid has no effect under non-inductive photoperiods, but increases flowering under inductive photoperiods in *Chenopodium* and *Pharbitis nil*. It is suggested that the AbA can not substitute for photoinductive treatment but might be additive to initiate reproductive stages.

Salicylic acid and some other phenolic acids have also been shown to induce flowering in *Lemna* and several other plants. S.C. Maheshwari and his colleagues (1987) while studying flowering in the duckweed *Wolffiela hyalina* observed that the plants did not flower when grown on basal nutrient medium under long day or short day photoperiodic conditions. However, when salicylic acid was also included in the nutrient medium, it did flower under short day conditions. But most other plants do not have an obligate requirement of salicylic acid for flowering, it merely hastens the flowering. The salicylic acid has other morphogenetic effects also, and it has been suggested that it acts like a plant growth regulator.

Ascorbic acid is also known to induce flowering in some plants such as *Lemna, Trigonella* and *Brassica*.

Low Temperature Treatment—Vernalization: In many species, temperature has a profound effect on flowering. Some plants do not flower even under the inductive photoperiod conditions, and flower only when a cold temperature treatment is given to the plant. This "acquisition or acceleration of the ability to flower by a chilling treatment" has been termed as vernalization (Vernal = spring like). Perhaps the first observation of vernalization or acceleration of flowering by cold treatment was made by a Russian scientist T.D. Lysenko (famous for Lysenko genetics). He demonstrated that the winter varieties of wheat, rye and barley, could be planted in the spring to yield the crop at the same time, the summer varieties did, if their moist seeds were subjected to an extended period of temperature fairly near the freezing point. The vernalization can be demonstrated in biennials, which normally require a season of wintering before flowering. It must be realized however, that vernalization per se does not induce flowering, but merely prepares the plant for flowering. This is in contrast to the photoperiodic effect on flowering, when a correct photoperiod not only prepares the plant for flowering but also initiates flowering.

The vernalization effect can be best visualized by examining the effect of chilling on flowering in winter rye, *Secale cereale.* It has two strains; Spring and winter strain. The spring strain is an annual, flowering and fruiting in one growing season. The winter strain is a biennial; staying vegetative in the first growing season (fall and winter) and flowering and fruiting in the next season (summer). However, when the winter strain is vernalized, it can be planted in spring and induced to flower in only one season and in this manner resembles spring strain. After a prolonged treatment at cold temperature, it responds to a proper photoperiodic treatment and flowering is initiated. This requirement of low temperature can be satisfied at any point in the developmental history of the plant after germination.

For example, if a seed of winter strain is permitted to germinate and then exposed to low temperature (about 2 to 5°C), it behaves just as if it had gone through a cold winter after a year of growth. It flowers in the first growing season, if exposed to proper photoperiod. Thus, vernalization changes a biennial into an annual. It should also be realised that in rye, vernalization is not an absolute requirement for flowering, it just shortens the time to flower. However, in many other biennials, its requirement is absolute and they can not flower without vernalization.

Not all the species can be vernalized at seed stage. For example, *Hyoscyamus niger*, has to be at least 10 days old and in rosette stage, before it can be vernalized. Generally those species, which can be vernalized at seed stage are facultative cold requiring plants, whereas those which can be vernalized at the plant stage show an obligate chilling requirement. For the majority of the species, the most effective temperature for vernalization are just above freezing *i.e.*, 1 to 5°C, but temperatures ranging from -1 to -9°C are almost equally effective.

Site of Vernalization: In the seedlings and in mature plants, shoot spex seems to be the site of vernalization. This has been shown with localised low temperature treatment of different plant parts in celery, sugarbeet and chrysanthemums. Grafting experiments have also demonstrated that the shoot tip receives the vernalization stimulus, and the stimulus is translocated to the other parts of the plant. Once the tip is vernalized, the condition is transmitted to all other tissues formed subsequently, so that all other lateral tips are also vernalized. If the vernalized tip is removed, the secondary and tertiary lateral are developed, which are, in the vernalized state. In *Lunaria biennis*, however, it has been found that younger leaves are also capable of being vernalized, but older leaves, which have ceased growth, do not respond.

In plants, which can be vernalized at seed stage, plumule is the receptive tissue.

Physiological and Biochemical Changes during Vernalization: Since freezing is not essential to bring about the changes during vernalization, it is suggested that, physiological rather than purely physical processes are involved. This belief is further strengthened by the fact that cold treatment of rye grain is ineffective during anaerobic conditions. In cultured plants and tissues, a supply of sugar also seems to be essential during vernalization.

It has been suggested that some kind of chemical is formed in the vernalized tip, which either induces florigen or/and makes the tissue receptive for the florigen action. That a chemical is actually involved in vernalization is demonstrated by grafting experiments. If a growing tip is removed from a newly vernalized plant and transplanted to one that has not been vernalized, the new plant will flower, indicating that it became vernalized by the grafting of the tip. Such grafting experiments have been successful not between strains of the

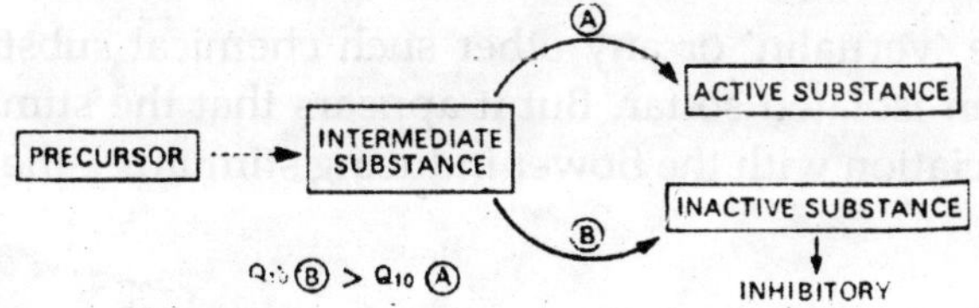

Hypothetical explanation of the apparent negative temperature coefficient of vernalization

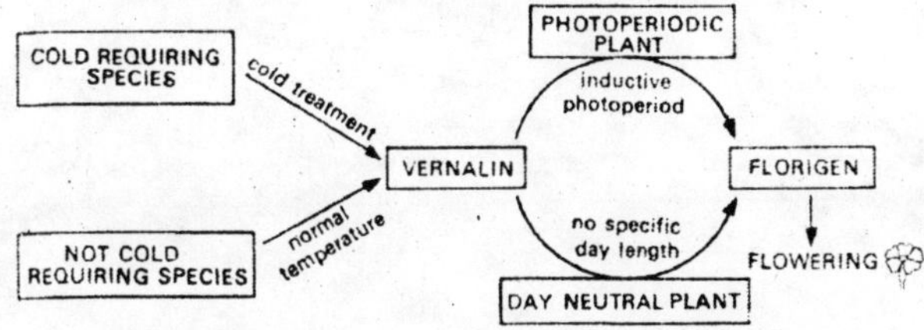

Hypothetical representation of the involvement of venalin and florigen in flowering.

same species, but also between different species and different genera. I. Melchers and A. Lang suggested that a transmissible flowering stimulus is formed as a result of chilling. They called this stimulus *vernalin*. Experiments have shown that this stimulus is not the same as florigen, as it can be formed under non-inductive photoperiods. It must be realized that the formation of vernalin is a very unusual process, because it has a negative coefficient of temperature.

It goes faster (or better) at lower temperatures. O.N. Purvis has hypothesized that there are two reactions involved in vernalization, in which one has a much higher temperature coefficient than the other. If the two reactions can be represented as A and B; the product of A is an active substance and that of B is inactive one. If the Q_{10} of reaction B is higher than that of A, then at higher temperature, an inactive substance will predominate, where as at low temperature active substance will be in abundance. If the formation of intermediate for these two reactions (A and B) is slow or if the product of B (inactive substance) is inhibitory, then the reaction will appear to have negative temperature coefficient.

The 'vernalin' or any other such chemical substance has not been isolated so far. But it appears that the stimulus acts in association with the flower inducing stimulus—the florigen.

5

Physiology of Plants

Water Absorption and Movement in Plant

- Water is the main source of plant and is essential for the maintenance of life, growth and development.

Various Methods

Imbibition Process

- This is the process of absorption of water by the particles of solid substances without forming a solution.
- Imbibition helps in the germination of seeds movement of water from cell to cell in the plant body and the ripening of seeds.

Diffusion Process

- This is the movement of molecules of gases, liquids and solutes from the higher concentration to the lower concentration area.
- The rate and direction of diffusion are influenced by the temperature and pressure.

Osmosis Process

- In the osmosis the movement of solvent molecules from a region of its high free energy to a region of its low free energy when the two are separated by a semipermeable membrane till a state of equilibrium is reached.
- Osmosis is of two types-endosmosis and exosmosis.
- Osmosis plays an important role in the absorption of water by the root-hairs, cell to cell movement of water, induction of turgidity, providing mechanical strength, induction of turgidity.
- Osmosis helps the plants in developing resistance to drought injury, movement of plant parts and opening and closing of stomata.

Plasmolysis Process

- This is the contraction and separation of protoplasm from cell walls due to exosmosis.
- The plasmolysis process is utilized in daily life in the preservation of meat, jellies and food stuffs and in preventing the growth of plants in the cracks of walls.
- The total amount of water present in the soil is called holard.
- The water available to the plant is called chesard.
- The water which cannot be absorbed is called echard.
- The symptoms appearing in a plant or the plant part due to the scarcity of water are known as wilting. Wilting may be temporary or permanent.
- Plants absorb water by two distinct mechanisms which independently operate. These mec hanisms are active absorption and passive absorption.
- Active absorption of water involves the utilization of metabolic energy. Water is absorbed by the activity of root hairs.

- Passive absorption of water accounts for about 98% of the total water uptake by a plant. This process does not involve the expenditure of metabolic energy.
- Root pressure, guttation and bleeding are example of active absorption in plants.
- Root pressure is the pressure developed in the xylem vessels as a result of metabolic activity of roots.
- Under conditions of water scarcity and low temperature the roots develop negative pressure.
- Guttation is the exudation of water drops from the edges or tips of the leaves.
- It occurs in the plants, growing under conditions favouring high root pressure and low transpiration, through the structure called hydathodes.

Bleeding

- It is the exudation of sap from the injured parts of the plants.
- The upward transport of water from the roots to the aerial parts of plants is called ascent of sap.

Cohesion

- Cohesion tension and transpiration pull theory as proposed by Dixon and Jolly.
- It is widely accepted theory to explain the mechanism of ascent of sap in plants.
- The loss of water in the form of vapours from the living tissues of aerial parts of the plants is called as transpiration.
- Approximately 98% of the water absorbed by the land plants is lost in transpiration.
- Three kinds of transpiration are distinguishable- stomatal, cuticular and lenticular.

- Stomatal transpiration accounts for about 90% of the total water loss from the plant by transpiration.
- The opening and closing of stomata is controled by the size and shape of guard cells resulting from the changes in their turgor.
- Transpiration is influenced by environmental factors such as-light, temperature, atmospheric humidity, wind velocity etc. and internal factor also influenced of transpiration.
- Transpiration is a necessary evil because of its advantages and disadvantages.
- Antitranspirants are the chemical substances that reduce the transpiration rate without effecting the other metabolic activities of the plants.

Exercise of Photosynthesis

"Photosynthesis is the process by which the green chlorophyll containing cells of the plants manufacture the carbon compounds with the help of inorganic raw material such as CO_2 and water in the presence of sunlight."

- The atmosphere contains only about 0.03% CO_2 by volume.
- The earth surface receives about 650×10^{21} cal of sunlight energy per year.
- It is estimated that only 1.2% of this total energy is utilized in photosynthesis.
- All green parts of the plants contain chloroplasts and leaves are the main sites of photosynthesis.
- Chloroplast pigments are of three types:

Chlorophylls

- This is green photosynthetic pigments.

Carotenoids

- This is yellow, orange or red coloured pigments and

absorb primarily the violet to blue regions of the spectrum.

Phycobilin

- According to the modern views, photosynthesis is fundamentally an oxidation-reduction process in which H_2O is oxidised to O_2 and CO_2 is reduced to carbohydrates and the radiant energy of the sun is stored in the form of chemical energy.
- According to the modern concept, photosynthesis is completed in two distinct steps namely light reaction and dark reaction.
- In the light reaction, ATP and $NADPH_2$ are generated and the water molecules undergo splitting resulting in the evolution of molecular oxygen.
- Photosystem-I is associated with both cyclic and non-cyclic photophosphorylation and reduction of NADP to $NADPH_2$.
- While photosystem II is concerned with only non-cyclic photophosphorylation and photo-oxidation of water resulting in the evolution of molecular oxygen.
- Dark reaction is also known as black man reaction in honour of the scientist who reported it first.
- The biochemical reaction of the "dark reaction phase" also known as Calvin cycle takes place in the stroma part of the chloroplast where all the necessary enzymes are located.
- Dark reaction primarily consists of three steps:
  (i) carboxylation
  (ii) glycolytic reversal and reduction
  (iii) regeneration of RuBP
- Photorespiration is the process of uptake of O_2 and production of CO_2 in light by photosynthesising tissues.
- Photorespiration is the common in C_3 plants.

- Photorespiration is also called as C_2 cycle because of 2-carbon intermediates.
- C_4 plants (Sugarcane, maize, sorghum etc.) Partially overcome the disadvantage of photorespiration.
- C_4 plants show a characteristic Kranz anatomy of the leaves.
- Parts of the plant such as the green tissues (leaves) which synthesis the carbohydrates are known as "Source".
- The parts of the plants such as non-green tissues where carbohydrates are utilized for maintenance growth and development are called as "Sink".

6

Plant Morphology

Morphological Characters

Morphological characters of plants have been used extensively both for producing classification and for diagnostic purposes and they are still indispensable to the taxonomist. However, external morphological study alone is not adequate and other branches of study are of considerable value in proper assessment of the systematic status of a taxon and its phylogeny. Recent researches have shown that contributions to systematics can come from any branch of botany. Vegetative anatomy, floral anatomy, palynology and embryology have played a significant role in plant taxonomy. The taxonomic contributions of cytogenetics have been enormous and chemotaxonomy has made an equally great impact on our ideas of classification and of phylogeny.

These sources of taxonomic evidence are considered briefly in the following pages.

External Morphology

The angiosperm classification is morphological bias as morphological characters have been extensively used both for producing classification and for diagnostic purpose. These characters are still indispensable and will continue to play an important role in plant taxonomy. This does not undermine the importance of other kinds of characters which have been helpful in their own way. The greatest advantage of morphological characters is that they can be easily observed in the field by a hand lens and does not need any elaborate laboratory facility.

The vegetative characters are as important in the identification of the majority of flowering plant as do the floral characters. It is true that vegetative characters show greater plasticity as compared with floral features and this makes their description more difficult. But all vegetative features are not difficult to assess and they have been profitably used by the taxonomists. Some non-traditional vegetative characters are discussed here (Davis and Heywood, 1963).

Habit, which concerns such features as size, branching, spread, density, rooting system, means of propagation and perennation and periodicity, is seldom treated adequately in taxonomic descriptions. Corner (1952) has remarked that space of trees which depends on the shape of their crowns (round and bushy, oblong or cylindric, umbrella-shaped or flat topped, etc.) is the significant factor in the recognition of trees. Characters of bark, such as colour, texture, thickness, fissuring, etc. are widely used in recognition of the species of *Pinus* and *Betula*.

The underground parts of the plant also provide valuable taxonomic characters. The form of the root structure is widely employed in the taxonomy of the genus *Ranunculus*. Species of *Aristolochia* can be delimited on the basis of the shape of the main root (globose, ovoid, elongated and cylindrical, fusiform or napiform).

In preference of floral features, leaf characters are widely employed for delimiting species of many genera such as *Ulmus* and *Betula*. Leaf pytix (rolling or folding of leaves) is of apparent taxonomic value in some woody Rosaceae. Leaf base and associated structures have been found to be of taxonomic value in Fabaceae. Stipular morphology is a useful character for keying out various species in genera like *Trifolium* and *Viola.*

Seedling morphology also provides characters of taxonomic value. They have proved of great help in the study of geographical variation in *Pinus contorta.*

The traditional basis for much taxonomic and descriptive work has been the characters derived from the inflorescence and flower. The students are familiar with such traditional characters. A detailed scrutiny of some non-traditional floral characters may provide surprising results. Distinctive features of adaxial petal appendages and the attachment of the petals to the rim of the cup warrant recognition of *Hespreolinon* (Linaceae) a distinct genus from *Linum* (Sharsmith, 1961). The degree of inflation of bracts and their shape clearly distinguishes species in the *Calystegia sepium silvatica* complex (Convolvulaceae). Similarly, in *Anthyllis vulneraria* complex two main groups are separated on the basis of calyx types (Cullen, 1962).

The configuration of floral disc and nectaries has proved to be of great diagnostic value in *Tamarix* and Brassicaceae.

Staminal appendages provide useful diagnostic characters in the species of *Alyssum* (Brassicaceae). The genus *Eucalyptus* has been classified on the basis of anther characters. Similarly, staminodes provide good specific diagnostic characters in the genus *Scrophularia.*

The pattern of papillae on stylar branches and the degree of pubescence of young ovaries contribute in the classification of Asteraceae.

The degree of branching of inflorescence proved useful in distinguishing various species of *Nepeta.*

The external characters of seeds are of great value in separating closely related genera and species. Markings of testa have been successfully used in separating species of *Chenopodium, Euphorbia, Hypericum,* etc.

The structure of the seed and the form and dimensions of the hypocotyl are widely used in agriculture for the recognition of races of economic plants such a *Brassica* species. Similarly, seeds of many weeds can be identified on the basis of their structure and several seed atlases have been prepared to allow the easy detection of contaminated seeds in crop material.

Vegetative Anatomy

Anatomical characters of the vegetative organs of flowering plants have been employed with great success to the solution of taxonomic problems and to the elucidation of phylogenetic relationships. They have also proved very helpful for the identification of fragmentary material (of economic importance) and of herbarium specimens which are not accompanied by flowers and fruits (Metcalfe, 1968). Carlquist's *Comparative Plant Anatomy* (1961) is an important source book of taxonomic and evolutionary application of anatomical data in angiosperms.

There are a large number of anatomical characters of systematic importance but, as pointed out by Metcalfe and Chalk. (1950), the systematic anatomist must rely on those characters only which are less plastic. They further suggested that conclusions supported by combination of characters are more reliable than those rest on a single character. Some important anatomical characters of well established taxonomic value are discussed in the following pages:

Trichomes: Trichomes have been employed very frequently for systematic comparisons because of their diversity, their universal presence in the angiosperms and their simple means of preparation. Several trichome types have been recognised. The two major categories of trichomes are, glandular and nonglandular and these are further subdivided on the basis of their gross form, cellular constitution, degree of branching, etc. Solereder (1908), Netolitzky (1932) and Metcalfe and Chalk (1950) have provided useful information on the structure, function and classification of trichomes.

Certain families can be recognised by the presence of distinctive type(s) of trichomes but they have proved more useful to taxonomists at generic and specific level. Phylogenetic trends have been recognised in trichomes within a given group. For example, Carlquist (1958 a, 1959 a, b) has shown that in subtribe Madinae of Heliantheae (Compositae) from simple glandular trichomes, capitate trichomes with multiseriated head have been derived and the latter have further given rise to hollow stalked glands, tack-shaped glands and sessile glands.

Trichomes form an excellent criterion on subgeneric and specific levels in *Rhododendron* (Cowan, 1950). They also contribute to ideas on relationship within the genus. Organographic distribution and types of trichomes provide useful distinguishing characters within Icacinaceae as shown by Heintzelman and Howard (1948). They supported segregation of *Ottoschulzia* from *Poraqueiba* on the basis of evidence from trichomes. Goodspeed (1954) has reported that types of trichomes and their distribution are correlated with subgeneric and specific distinction in *Nicotiana*. Shah and Kothari (1973) showed that the distribution of hairs in the tribe Vicieae (Fabaceae) is of considerable help in classifying various genera.

Occurrence of sessile glandular trichomes in Sparganiaceae and Typhaceae provides a link between these families (Solereder and Meyr, 1933). On the basis of a detailed study of the morphology and ontogeny of trichomes in the tribe Heliantheae (Asteraceae), Ramayya (1962, 1969) finds support to the view that this is the most primitive tribe of Asteraceae. On the basis of structure and ontogeny of trichomes, Inamdar (1967) confirms the position of *Nyctanthes* in Oleaceae.

Stomata: Vesque (1889) proposed first classification of stomata and he recognised four morphological types, named after the families in which they were first observed. Later, Metcalfe and Chalk (1950) also classified the stomata on the same plan with some modification in terminology. Pant (1965) recognised 10 stomatal types mainly on the basis of their ontogeny. Van Cotthem (1970) proposed an elaborate classification based on the number and position of subsidiaries and on ontogeny.

Ontogenetical pattern of stomata, their morphology and spatial relationship to the neighbouring cells and the number and arrangement of subsidiary cells are the characteristics of stomata which may be helpful in systematic and phylogenetic considerations. Stomata, along with other epidermal features, have also been employed in the identification of fragmentary materials in the pharmacognostical studies (Krishnamurthy and Kannabiran, 1970).

Prat (1932) has shown the utility of stomata in the systematics of the Gramineae and he suggested the removal of the genus *Jouvea* from the Hordeae. According to Watson (1962), the pattern of stomatal distribution is an important taxonomic criterion in Epacridaceae as also structural peculiarities of stomata. For example, he observed that in the tribe Styphelieae stomata are present on the abaxial surface of the sepals whereas they are usually adaxial in the others. He further observed that while the family is characterized by the presence of anomocytic stomata, three genera possess paracytic stomata.

Several distributional and structural features of the stomata, such as their shape and size, degree and pattern of thickening of the walls of guard cells, distribution of cuticular thickenings and their level in relation to the epidermis are of considerable taxonomic significance in the Combretaceae and have been used at the specific level (Stace, 1965).

On the basis of structure and the development of stomata in Rubiaceae, Pant and Mehra (1965) and Bir *et al.* (1971) substantiate the classification of Verdcourt (1958) who divided the family into three subfamilies, Rubioideae, Cinchonoideae and Guettaroideae. Exhaustive studies of Guyot (1966, 1971) on the stomata of the Umbelliferae support the classification of Cerceau-Larrival (1962) based on pollen morphology. Shah (1968) and Kothari and Shah (1974 a, b, 1975) discussed the taxonomic and phylogenetic significance of stomata in the

Papilionaceae. While primitive types of stomata (anomo-perigenous and aniso-mesoperigenous) are more common in the tribes Sophoreae, Podalyrieae and Gensiteae, advanced types occur more frequently in Dalbergieae, Phaseoleae and some Hedysareae.

The ontogenetic studies of the stomata of Cucurbitaceae and Caricaceae suggest close affinity between these families as both possess aniso-mesoperigenous and aperigenous stomata. The regrouping of the families Alangiaceae and Cornaceae in the order Cornales and Araliaceae and Umbelliferae in the Umbellales by Cronquist (1968) is substantiated by the stomatal studies. The former two families possess aperigenous stomata while the latter in addition also have anisocytic and paramesogenous stomata (see Kannabiran and Krishnamurthy, 1979).

Studies of Paliwal (1966) on the ontogeny of stomata indicate that in the presence of diacytic stomata and syndetochelic mode of development *Elytraria* resembles other members of the Acanthaceae. These features are in contrast to those of the Scrophulariaceae where stomata are anomocytic and haplochelic and without subsidiary cells. This supports the retention of *Elytraria* in Acanthaceae and negates the suggestion of Bremekamp (1953) to transfer it to Scrophulariaceae. Similarly, retention of *Nyctanthes* in Oleaceae is also substantiated on the basis of ontogenetic study of stomata (Inamdar, 1968). Like other Oleaceae, *Nyctanthes* possesses tetramesoperigenous and occasionally aperigenous stomata in contrast to diamesogenous stomata of the Verbenaceae to which transfer of *Nyctanthes* is suggested.

Epidermis: The shape of the epidermal cells, thickness and characteristics of their wall, nature of sculpturing on their walls as seen in surface view and inclusions of epidermal cells provide useful taxonomic criteria. Divisions of epidermal cells, although of restricted and sporadic occurrence, are also of some diagnostic value in certain families, such as Piperaceae.

Cuticular relief is found to be a diagnostic character in ashes (Munz and Laudermilk, 1949). A characteristic feature of Winteraceae is occlusion of stomata by an unidentified substance (Bailey and Nast, 1944). Sclerification of the walls of epidermal cells is a feature of taxonomic importance in the tribe Mutisieae of Compositae (Carlquist, 1958 b) and extremely narrow cells are characteristic of the epidermis of Stylidiaceae (Mildbraed, 1908).

The importance of papillate epidermal cells in the systematics of Gramineae has been emphasised by Prat (1932). Similarly, the shape of silicified cells and suberized cells and their specific relationship are of diagnostic value in certain Gramineae (Prat, 1932). Different types and sizes of "light receptors" in *Mesembryanthemum* provide useful specific criteria (Kean, 1931). Bladder like epidermal cells are characteristics of some families (Metcalfe and Chalk, 1950).

Details of the silica-body distribution as seen in the surface view of the leaf also provide good taxonomic criteria in Cyperaceae, *e.g.*, wedge-shaped silica bodies are restricted to *Mesomelaena, Neesenbeckia, Ptilianthelium, Scirpodendron* and *Thoracostachyum*, bridge-shaped bodies to *Mapania;* spherical and hemispherical warty bodies to *Acriulars, Bisboeckelera* and *Scleria;* spherical and hemispherical echinulate to *Rhynchospora* and hemispherical 'smooth bodies to *Lophoschoenus* (Metcalfe, 1968).

Ahmad (1974) has shown that epidermal characters of different species of *Hemigraphis* and *Strobilanthes* are sufficiently pronounced as to justify their use in making distinctions at specific level. He (Ahmad, 1976) also justifies the treatment of *Asteracantha longifolia* as *Hygrophila auriculata* by Heine (1962) on the basis of his study of foliar epidermis.

Singh *et al.* (1974) supported the retention of *Berberis* and *Mahonia* as separate genera as the two differ in several epidermal features. On the basis of epidermal features such as the outline of epidermal cells in surface view, structure of stomata and type of hairs, Jain and Singh (1974 a) and Singh and Jain (1975) have drawn keys to distinguish various species of *Pyrus* and *Prunus.* Similarly, Jain and Singh (1974 b) have been able to distinguish Himalayan species of oaks on the basis of epidermal features.

Leaf Architecture: The most comprehensive system of descriptive terminology of gross leaf form and venation pattern of dicotyledonous leaf has been published by Hickey (1973). Data obtained from leaf architectural features, particularly venation pattern have been successfully utilized for systematic and evolutionary considerations. Foster (1959) suggested that open dichotomous venation of the ranalian genus *Kingdonia* may be primitive within angiosperms.

Carlquist (1959 c) reported that minor venation ,pattern in certain Hawaiian species of Compositae provides taxonomic and evolutionary clues. Dede (1962) described the foliar venation pattern in the Rutaceae and deviced a key for the identification of various species on the basis of these characters. Handro (1964, 1967) has also indicated the possibility of venation and features of leaf anatomy in the distinction of genera and species in the Amaranthaceae. Tucker (1964) demonstrated the utility of veinlet endings in the diagnosis of the two species of *Liriodendron, L. chinense* and *L. tulipifera.* Banerji and Das (1972) have shown that minor venation pattern is useful in the distinction of the Indian species of *Acer.* The size of areoles, number of terminations per areole and characteristics of branches provide useful taxonomic characters.

Singh *et al.* (1978) described leaf architecture in Berberidaceae and discussed its bearing on the circumscription of the family. They concluded that in leaf architecture, *Holboellia* and *Podophyllum* stand apart from other taxa of the family and this supports their removal to separate families. Differences in the leaf architecture of *Berberis* (presence of simple tufted leaves, mid-vein elevated on the adaxial side, presence of simple intersecondary veins, straight or slightly curved primary vein, more acute secondaries in the basal part of lamina and presence of vein endings) and *Mahonia* (presence of pinnately compound leaves, mid-vein grooved on adaxial side, absence of intersecondary veins, zig-zag primary vein, and absence of vein endings) justify the recognition of these two taxa as separate genera which were once included in *Berberis*.

Hickey and Wolfe (1975) provided first systematic summary of dicot leaf architectural features and they demonstrated that a number of lower order leaf architectural features, including leaf organisation, configuration of first three vein orders and characteristics of leaf margin are significant systematic indicators within dicotyledons. Leaf data support Takhtajan's (1969) assignment of the Euphorbiales to the Dilleniidae; Cronquist's (1968) assignment of the Lecythidaceae and related families to the Dilleniidae, rather than to the Rosidae as has been done by Takhtajan and erection of two orders for Takhtajan's order Cornales. Leaf architectural patterns also suggest separation of the Asteridae into two groups, reassignment of the Celastrales and Myrtales to the Dilleniidae and of the Juglandales to the Rosidae.

Leaf Anatomy: Carlquist (1961) has rightly stated that "the leaf is perhaps anatomically most varied organ of angiosperms and its anatomical variations often concur closely with generic and specific and occasionally familial lines". The leaf possesses many systematically valuable histological features such as structure of mesophyll, and vein sheaths, presence or absence of hydathodes, foliar nectaries and stomatal crypt and time of formation, suberization and location of abscission layer.

The leaf anatomy is one of the most reliable characters useful in grass systematics. This was realized as early as 1875 by Duval-Jouve and since then has been emphasised by several workers. Two distinct types were recognised by Avdulov (1931) on the basis of arrangement of internal tissues in grass leaf blade. The festucoid type has a distinct thick walled mesotome (endodermal) sheath around the vascular bundle and a very irregular arrangement of mesophyll and the panicoid type is characterized by the presence or absence of a mesotome sheath and more or less radial arrangement of mesophyll. Brown

(1958) recognised six types of tissue arrangement in grass leaf, that he correlated with taxonomic groups. He has also discussed evolutionary tendencies in the various groups of grasses on the basis of leaf anatomy.

The distribution pattern of sclerenchyma in the leaves of *Festuca* is of great value in distinguishing various species that are otherwise difficult to be distinguished from one another. It is also of very great value in Cyperaceae for separating species of *Carex* and is also of value at the generic level (Metcalfe, 1968). Ayensu (1974) has shown that *Vellozia* and *Barbacenia* (Velloziaceae) can be distinguished from one another by the form of sclerenchyma in the leaves.

Anatomical investigation of Cutler (1965) on South American genus *Thurnia* is an instance of leaf anatomy giving a positive taxonomic lead. He found several remarkable anatomical features in *Thurnia,* such as inverted bundles, a unique type in the leaf. Thus, this provides further support to the view that *Thurnia* should be placed to a family of its own, the Thurniaceae separate from the Juncaceae and Rapateaceae to which it was previously assigned.

Of the many studies on systematic implication of leaf anatomy, mention may also be made of Wunderlich (1950) on *Agave,* Hagerup (1953) on Ericaceae, Morley (1953 a, b) on *Mouriri,* Fahr (1954) on Xanthorrhoeaceae and Tomlinson (1956, 1959 a) on Zingiberaceae and Musaceae.

Smithson (1956) has shown that Flagellariaceae resemble to the Gramineae in their foliar anatomical structure. Leaf anatomy has proved useful in establishing correct systematic position of *Dianthonia* and its allies (De Wet, 1954). Govindarajalu (1969) was able to formulate a key to distinguish various species of *Cyperus* on the basis of leaf anatomy.

Petiole Anatomy: Vascular system of dicotyledonous petiole, as seen in transverse sections through the distal end, is of considerable taxonomic importance. Several types of vascular structures of petiole are illustrated by Metcalfe and

Chalk (1950). Many taxa have special type of vascular structure, *e.g.*, in *Anemone vitifolia* petiole has scattered vascular bundles.

Soladoye (1982) examined the petiole anatomy of 64 species of *Baphia* (Leguminosae). He observed certain anatomical characters of the petiole, such as petiole outline, number of layers of parenchyma cells in the cortex, distribution of perivascular fibres and vascular system which provide supporting evidence to be used in infra-generic classification of the genus. They correspond to systematic groupings identified on other criteria.

Vascular structure of petiole is of much diagnostic significance in the genera *Phlomis* and *Eremostachys* (Labiatae) (Azizian and Cutler, 1982). Two distinct vascular bundle arrangements were observed. The first, with 1-2 median arcs, noted in *Phlomis* section *Phlomis* and the species of *Eremostachys*. The second type, characterised by the occurrence of numerous separate bundles, often in a complete cylinder, occurs in *Phlomis* section *Phlomis*, but is absent from *Eremostachys*. This confirms that within *Phlomis* two distinct groups can be recognised and that *Phlomis* and *Eremostachys* are closely related.

Nodal Anatomy: Sinnott (1914) recognised three fundamental types of nodal anatomy in dicotyledons, unilacunar, trilacunar and multilacunar. He put forward views concerning the phylogeny of the angiosperms on the basis of nodal organization. He believed that trilacunar node is primitive and that unilacunar and multilacunar, which are derived from trilacunar type to be more advanced. Marsden and Bailey (1955) discovered the fourth type of nodal anatomy, unilacunar two trace and this led to the revision of concepts of nodal evolution. Now unilacunar two trace node is considered basic type for angiosperms as this type is wide spread in vascular plants (other than angiosperms) and in majority of cases cotyledonary nodes have even number of traces.

The number of gaps may be characteristic for a single family or even for a order, as in the Centrospermales where the node is unilacunar in the whole order or there may be much variation within a taxon as in Ranunculaceae (Sinnott, 1914) or even within a single plant (Bailey, 1956). A comparative study of nodal anatomy may be valuable for showing relationships or distinctness of genera or even species (Sinnott and Bailey, 1914).

An apparently phylogenetic sequence seen in the mature unilacunar node of the Monimiaceae *(sensu lato)* involves a series commencing with two traces *(Austrobaileya, Trimenia, Piptocalyx)*, followed by 3, 5 or 7 traces *(Hortonia, Anthobembix, Mollinedia)* and finally ending with a single broad arc-shaped leaf trace Siparuneae (Canright, 1955).

On the basis of nodal structure, the subfamily Icacinoideae of the family Icacinaceae may be divided into two distinct sections, one section characterized by having trilacunar nodes and the other by having unilacunar nodes. One section of the tribe Icacineae with unilacunar node appears to be somewhat transitional in form between non-scandent trilacunar Icacineae and the unilacunar Iodeae, Sarcostigmateae and Phytocreneae in which twinning or climbing habit is dominant (Bailey and Howard, 1941).

Histology of the Stem: Internal structure of the primary stem may provide useful diagnostic characters in many cases. The distribution and abundance of collenchyma and pattern of collenchyma thickenings are important to the systematists. Transformation of ground tissue cells of cortex into transfusion cells forms a useful taxonomic feature in some taxa such as *Casuarina* (Metcalfe and Chalk, 1950). Distribution of fibers may be systematically useful as evident in the species of *Genista* (Pellegrin, 1908). Stem endodermis shows much variation in the structure of their cells and it may provide systematic criterion in families like Piperaceae, Asteraceae and Lamiacae. Features of stem pith may also be systematically useful. For example, species of *Dubantia* and *Fitchia* may be distinguished on the basis of anatomical differences in pith (Carlquist, 1961).

Stace (1970) has been able to distinguish most British species of the subgenus *Genuini* of *Juncus* on the basis of histology of the stem, particularly size and shape of sclerenchyma girders, degree of elevation of stem ridges and characteristics of pith. He could also trace parents of several hybrids on the basis of these characters.

Vascular core of the stem is taxonomically useful in several cases. Occurrence of bicollateral vascular bundles in two alternate rings is a characteristic feature of the stem of Cucurbitaceae. The occurrence of cortical and medullary bundles, restricted to some families such as Nyctaginaceae, Amaranthaceae, Chenopodiaceae, are features of diagnostic value.

Ayensu (1970) has shown that the two species of *Dioscorea, D. cayenensis* and *D. rotundata* which are difficult to distinguish on exomorphic grounds can be separated on the basis of varied arrangement of vascular bundles in the stem.

Various species of the genus *Potamogeton* can be distinguished on the basis of vascular core of the stem (Ogden, 1943). The species with single bundle type are perhaps highly advanced.

Anomalous secondary thickenings occur in several families of dicotyledons, and Bailey and Howard (1941) have shown their usefulness as taxonomic criteria. However, the families in which anomalous thickenings occur are seldom closely related to one another.

Sclereids: Sclereids are specialized cells found in almost all the parts of plant body. They serve in various ways to aid taxonomy and form important diagnostic tool in many taxa. They have proved useful in verifying the naturalness of taxonomic groups and in making internal rearrangement (Bokhari and Wendelbo, 1976; Rao, 1979).

Attempts have been made to utilize the sclereids in systematic studies at the level of the family (*e.g.*, Connaraceae, Goupiaceae, Hamamelidaceae, Limoniaceae, Nymphaeaceae, Oleaceae, Theaceae, Sonneratiaceae and Thymelaeaceae), sections (*e.g.*, Abutoideae of Ericaceae, Nelsonieae of

Acanthaceae and Monesterioideae of Araceae) and genus (*e.g.*, *Boronia, Boronella, Camellia, Capparis, Cyrtandra, Fagraea, Garrya, Jasmirnum* and *Rhizophora)* (see Rao, 1980).

The presence of elongate sclereids in the epidermis of ovules of Cynareae has been helpful in defining that tribe (Carlquist, 1961). The four sections of *Mouriri* have distinctive sclereid types (Foster, 1946). On the basis of variations in morphology and distribution of sclereids, Barua and Wight (1958) analyzed *Camellia* species. The taxonomic utility of sclereids has also been demonstrated by Rao (1957) in *Memecylon* (Melastomataceae) and by Temlinson (1959 b) in palms.

Cellular Contents: Many types of chemical deposits such as plastids, starches, proteins, tannins and crystals and other mineral deposits occur in plant tissues. Their distribution pattern and variety are of considerable diagnostic importance.

The distribution pattern of the cells containing chloroplasts and other types of plastids is of systematic value. The microscopical characters of starch grains such as their shape and size and their appearance in polaroid light are highly distinctive. Reichert (1913) has shown that a careful study of starch grains of particular groups may reveal generic, specific and even varietal criteria.

Comparisons of solid protein depositions have potentialities for systematic use. Some Cactaceae have characteristic protein bodies (Metcalfe and Chalk, 1950). Three species of *Laportea* may be distinguished on the basis of "albuminoids" from laticifers (Guerin, 1923).

Deposition of silica, calcium oxalate, calcium carbonate and other chemical substances is characteristic of certain taxa and they have been found of systematic value. In grasses and sedges silica bodies and silica cells have been found to be of systematic value. Carlquist (1961) has shown that variations in size, number per cell and spininess of silica bodies provide systematic characteristics in certain Rapateaceae. Epidermal cells containing large silica bodies (these cells termed as stegmata) are of great systematic use both at generic and specific level in Zingiberaceae, Musaceae and Arecaceae (Tomlinson, 1956, 1959 a, b).

Various types of crystals of calcium oxalate occur in plant body of many taxa. Chartschenko (1932) has shown that the species of *Allium* can be distinguished by means of oxalate crystals and Rothert (1900) found them useful in comparing species of *Eichhornia.* In some cases cystoliths are also helpful in linking related families, *e.g.,* Moraceae, Urticaceae and Cannabinaceae (Carlquist, 1961).

There are distinctive types of tanniniferous cells and three types are recognised in *Eupomatia.* The presence and absence of tanniniferous cells in the root cortex of related families Rapateaceae and Xyridaceae respectively is a useful systematic criterion. The presence of ethernal oil cells is the most important criterion that may be used to unite the heterogeneous order Ranales (Carlquist, 1961).

Wood Anatomy in Relation to Taxonomy: The data from wood anatomy are of considerable taxonomic and phylogenetic significance. This branch of study has provided additional evidences which have proved invaluable in the solution of taxonomic problems. A number of such instances are cited by Tippo (1946, 1950), Carlquist (1961) and others. There are a number of features of xylem which show major trends of evolution in angiosperms. Emphasizing the importance of xylem evolution Carlquist (1961) states that "the well established trends of xylem evolution should be taken into account by any systematist interested in formulating relationships or evolutionary status of angiosperm taxa." Some important features are listed below.

Vessel Elements: The vessel elements are considered phylogenetic derivatives of tracheids as primitive vessel elements show great morphological similarity to tracheids. The vessels with great length and comparatively narrow width, long overlapping end walls, scalariform pitting on lateral walls, scalariform perforation plates with many bordered perforations and angular walls in transectional view would be primitive. The evolution of the vessel element has occurred along the following lines: (i) There has been a phylogenetic decrease in vessel element length corresponding to advancement, (ii) There has been alteration of end wall from oblique to nearly transverse angle, (iii) There have been loss of borders and the decrease of bars on the perforation plate. By gradual narrowing of vessel the perforation plate may be lost resulting in a vascular tracheid. Examples of this are found in advanced families such as Cactaceae and Asteraceae. (iv) Lateral walls of vessels show evolutionary series in pitting type from scalariform to transitional to opposite to alternate. (v) There has been a transition from vessels with angular in outline (as seen in transverse section) to vessels with nearly circular outline.

Vessel abundance (number of vessels per square millimeter of area) also offers a characteristic of taxonomic significance.

Solitary vessels are considered as primitive while the various aggregate groupings, such as pore multiples, pore clusters and pore chains as advanced expression.

Distribution of vessels in transection is a taxonomically significant feature. The diffuse-porous woods are considered to be more primitive than the ring-porous woods. Sculpturing on vessel walls is a useful diagnostic criterion. The presence and absence of tyloses and presence of visible pits in tyloses also afford taxonomic criteria.

Vascular Rays: The various features of vascular rays offer useful taxonomic criteria. These include ray abundance, dimensions of rays as seen in tangential sections, width and cellular composition of rays, degree of wall thickness of ray cells and pitting in ray parenchyma cells. Heterogeneous rays (rays with both vertically and radially elongated cells) are primitive than the homogeneous rays (rays with all cells radially elongated).

Axial Parenchyma: The distribution and characteristics of the cells of axial parenchyma are the most conspicuous features of the secondary wood. The parenchyma cells are variously distributed but they fall into two main groups: (i) apotracheal (parenchyma distributed without any specific relation to vessels) and (ii) paratracheal (parenchyma distributed in close association with vessels).

Axial parenchyma is considered to have been evolved in accordance with the major trends of xylem specialization. Absence of parenchyma is primitive at least in some cases, *e.g.*, Winteraceae where the wood is also devoid of vessels. The diffuse arrangement of parenchyma cells is more primitive than the diffuse-in-aggregates. The various aggregate arrangements such as apotracheal banded parenchyma and the various paratracheal types represent advanced conditions. The length and width of parenchyma cells and thickness and lignification of their walls also provide important diagnostic features.

Storied Wood: The plane of divisions of cambial initials determines the storied structure of wood. Storied woods have short tracheary elements and other advanced features. The woods with storied structures are thus considered advanced over the ones with non-stratified cells.

Other Characters: Certain other characters such as occurrence of laticifers in secondary xylem, secretions of gums, resins and other substances in wood and types and location of crystals in woods also provide valuable taxonomic criteria.

Some Examples of the Value of Wood Anatomy in Taxonomy

Magnoliales and Amentiferae: In Englerian system of classification the Amentiferae are considered as least specialized whereas most of the recent phylogenetists including Hutchinson, Cronquist and Takhtajan consider the Magnoliales as the primitive group. The wood anatomy supports the latter contention, as the Magnoliales show several primitive anatomical characters. They include some vesselless genera and others have scalariform perforation plates exclusively. The vessels are long with small diameter, oblique end walls and scalariform inter-vascular pitting. The wood is diffuse porous and the pores show usually solitary arrangement. Several taxa have heterogeneous rays and diffuse parenchyma.

On the other hand, the Amentiferae are highly specialized in the structure of their wood. They have ring porous wood, simple perforation plates, short and round vessels with opposite or alternate pitting, and homogeneous rays. This rejects the concept that the Amentiferae are a primitive group of angiosperms.

The anatomical studies also outline new natural relationships for the taxa once included in the Amentiferae. Moseley (1948) demonstrates affinities between Casuarinaceae and Hamamelidaceae. Fagaceae are considered to have been derived from rosalean ancestry (Hall, 1952), and there are similarities between Julianiaceae and Burseraceae or Anacardiaceae. Tippo (1938) and Moseley (1948) delineate affinities between Urticales and Hamamelidales. This clearly demonstrates that the Amentiferae are an unnatural group.

Ranales: Studies on anatomy of wood of Ranales revealed great diversity indicating that they are an unnatural order. On the one hand, there are primitively vesselless taxa such as *Trochodendron, Tetracentron, Sarcandra* and Winteraceae and on the other hand, the woods of Himatandraceae and Magnoliaceae show a wide range of specialization. The wood in scandent Ranales such as *Austrobaileya, Illicium, Schizandra* and *Kadsura* although primitive, is modified with relation to the twinning habit.

Rhoipteleaceae: The family Rhoipteleaceae has been variously placed in Urticales and in Juglandales. The anatomical studies of Withner (1941) concluded that they belong to Juglandales. Scalariform perforation plates are also present in Rhoipteleaceae as in Juglandales but they are absent in Urticales.

Eucommiaceae: The monogeneric family Eucommiaceae has been assigned to Hamamelidales or to Urticales. The wood anatomy supports the latter treatment as both Eucommiaceae and Urticaceae have latex and vessel elements are with simple perforation plates. These characters are, however, absent in Hamamelidales. Thus, according to Tippo (1940) the correct place of Eucommiaceae is near Ulmaceae.

Burseraceae and Allied Families: On the basis of wood anatomy the families Burseraceae, Meliaceae, Sapindaceae, Rutaceae, Simarubaceae and Anacardiaceae are considered to form a phyletic unit. This does not support the treatment of Engler who placed some of these families in the Geraniales and the others in the Sapindales. On the other hand wood anatomy lends support to the Hutchinson's treatment of these families.

Parietales: The anatomical evidence suggests that Parietales of Engler is a heterogeneous assemblage and Wettstein's division of this order into Parietales and Guttiferales is justified.

The studies on wood anatomy favour that *Degeneria* should be placed in a separate family, Degeneriaceae. Magnoliaceae, Degeneriaceae and Himatandraceae are closely related families. *Illicium* which has vessels, should be removed from the Winteraceae which has no vessels, and *Euptelea* should not be associated with *Trochodendron* and *Tetracentron*.

The wood anatomy may also be helpful in assigning the correct position of disputed genera. For example, studies of Stern and Brizicky (1958) have shown that the correct position of *Heteropyxis* is in the Myrtaceae.

Floral Anatomy in Relation to Taxonomy: The course and distribution of vascular bundles within the receptacle and floral parts have proved to be of much help to systematists in several cases particularly in ranking taxa of higher order such as genera and families. The role of floral anatomy in relation to taxonomy has been discussed in some details by Eames (1935), Puri (1952, 1958, 1962), and Murty and Puri (1980).

Typical vascular ground plan of the flower shows that the sepals receive three distinct traces, petals and stamens a single trace, and carpels three traces each. While discussing the principles of floral variations that affect the vasculature of flower, Puri (1952, 1962) has pointed out that during specialization and evolution of the flower the vascular plan undergoes considerable modification through (i) Reduction, (ii) Amplification, (iii) Cohesion, and (iv) Adnation of vascular bundles. These four principles have thus provided some useful data to the systematists.

Let us now cite a few instances where floral anatomy has contributed valuable data in concerning some disputed genera and determining the relationship of various taxa.

Paeonia: The genus *Paeonia* has been included in the family Ranunculaceae by the earlier systematists but subsequently it was removed from the Ranunculaceae and was placed in a family of its own, the Paeoniaceae. The floral anatomy also supports such a treatment. While in other genera of the Ranunculaceae the sepals receive three and petals one trace each, in *Paeonia* they receive few to several traces. In *Paeonia* the stamens are supplied by a few large trunk bundles, whereas in all other genera stamen traces are derived independently. The development of stamens is centrifugal in *Paeonia* in contrast to centripetal development in other genera of the Ranunculaceae. The anatomy of the carpel is also unlike that of the other genera. While carpels are three-trace in the other genera, they receive several traces in *Paeonia*. These differences justify removal of *Paeonia* from the Ranunculaceae.

Trapa: The genus *Trapa* which was included in the family Onagraceae, later given the rank of a separate family (Engler and Diels, 1938; Pulle, 1938). This separation is also justified on grounds of vascular anatomy. The vascular plan of the flower of *Trapa* is different from the other genera of the Onagraceae. Furthermore, the inferior ovary of *Trapa* has been considered to be receptacular due to down-turning of the receptacular bundles, whereas in the other genera of the Onagraceae it has been regarded as appendicular in the fusion of the bundles present in the same radii.

Cyrtandromoea: While discussing the classification of the Gesneriaceae, Burtt (1965) concluded that *Cyrtandromoea* has no close affinity with other members of the Gesneriaceae and that it should be transferred to the Scrophulariaceae with which it has better resemblance. Recently, Singh and Jain (1978) have also supported this treatment on the basis of floral anatomy. The sepal laterals arise conjointly with the petal traces in the Gesneriaceae while they emerge independently in *Cyrtandromoea.* The carpel *of Cyrtandromoea* has several lateral bundles but they are absent in the Gesneriaceae or when present they are only two in number. In the Gesneriaçeae the carpel laterals also contribute to the ovular supply in addition to the lateral strands. On the other hand, in the independent emergence of the sepal lateral traces, the presence of several lateral traces in the carpels, presence of compound ventral strands, as well as a bilocular ovary and in the absence of disc, *Cyrtandromoea* shows resemblance with the Scrophulariaceae.

Lilaea: There has been some difference of opinion regarding the systematic position of *Lilaea.* It was included earlier in the family Scheuchzeriaceae but subsequently segregated to a separate family, the Lilaeaceae. The latter view has also been concurred by Hutchinson (1959). The floral structures of *Lilaea* show many dissimilarities from other genera of the Scheuchzeriaceae. The gynoecium of *Lilaea* also differs from that of other members in the vascular supply and the number of ovules. This justifies the removal of *Lilaea* from the Scheuchzeriaceae.

Ruppia and Zanichellia: The two genera, *Ruppia* and *Zanichellia,* traditionally included in the family Potamogetonaceae along with *Potamogeton* were removed to separate families Ruppiaceae and Zanichelliaceae respectively by Hutchinson (1959). The study of Singh (1964) on floral anatomy gives no support to this treatment. In *Potamogeton,* the inflorescence is many flowered spike, whereas the number of flowers in an inflorescence is reduced to two in *Ruppia,* and in *Zanichellia* flowers are solitary. While the flowers of *Potamogeton* and *Ruppia* are bisexual, they are unisexual in *Zanichellia.* The perianth segment of *Potamogeton* is large and vascularized; that of *Ruppia* is reduced to a small outgrowth which is nonvascular and in *Zanichellia* perianth is entirely absent. A reduction can also be seen in the vascular supply of carpels of these three genera. All these characters show stages in reduction in the three genera and they seem to argue for keeping three genera together rather than separating them.

Hydrocotyle Asiatica: Urban transferred *Hydrocotyle asiatica* L. to the genus *Centella (C. asiatica* L. Urban). This has been confirmed by Mittal (1955) on the basis of the floral anatomy. While the inflorescence of *Hydrocotyle* is a umbellose raceme, in *H. asiatica* it is a cyme as in *Centella.* In the other species of *Hydrocotyle,* the ovular traces are derived from the placental strands, whereas in *H. asiatica* they receive their vascular supply from the alternate bundles in each carpel as in *Centella.*

Subfamilies and Tribes: In the same way the vascular anatomy of the flower has provided many additional evidences which have been used in the treatment of subfamilies and tribes in several families such as Ranunculaceae, Rutaceae, Fabaceae, Gentianaceae and Boraginaceae, and as an aid in generic limitation in others, *e.g.*, Aceraceae, Cornaceae, Ericaceae and Liliaceae.

Interrelationships and Affinities of Families: Floral anatomy has also provided valuable data in determining the interrelationships and the affinities of several higher taxa of family rank.

Anatomical study of the flower of Papaveraceae, Capparidaceae, Brassicaceae and Moringaceae shows that the gynoecial make up of these families is essentially on the same plan. In all these families although the placentation is parietal, the placental strands which occur on the inner side of the secondary marginals are inversely oriented with reference to the floral axis. This supports the inclusion of the Moringaceae in the Rhoeadales. This anatomical peculiarity has also been reported in the Passifloraceae and the Cucurbitaceae and thus strengthen the affinities between the Moringaceae and the Passifloraceae (Puri, 1947, 1954).

Most of the authors agree that the families Butomaceae and Alismaceae are closely related and hence Bentham and Hooker (1862-1883), Rendle (1930), etc. have included the taxa of both the families under a single family Alismaceae. Engler and Prantl (1889) have placed the two families in the sub-order Butomideae. Hutchinson (1959), however, placed the Butomaceae along with Hydrocharitaceae under the Butomales and the Alismaceae in a separate order the Alismatales. The studies on floral anatomy (Singh, 1964) also suggest resemblance between the Butomaceae and the Hydrocharitaceae. The ovary is inferior in the Hydrocharitaceae while the Butomaceae show a tendency towards epigyny. The vasculature of the gynoecium of the Butomaceae is very similar to that of many Hydrocharitaceae. On the other hand, the vascular anatomy of the flower of the Butomaceae differs considerably from that of the Alismaceae.

The floral vasculature of Menispermaceae, Calycanthaceae and Annonaceae suggests that these families have originated from the Ranunculaceae.

Vascular ground plan of the flower in general and the course of the placental bundles in particular in the Caryophyllaceae and the Polemoniaceae suggest that the latter have been derived from the caryophyllaceous stock.

The floral anatomy of the Solanaceae and the Scrophulariaceae suggests close affinity between the two families in the uniformity of vasculature of the gynoecium and in the presence of anatomically parietal placentation. Takhtajan (1969) has also envisaged resemblances between the two families and included them in his order Scrophulariales.

Phylogenetically Cyperaceae and Gramineae have been allied closely and are placed together in the same order in earlier systems of classification but Hutchinson placed them in separate orders, the Cyperales and the Graminales respectively. On anatomical evidence the single basal ovule of the Cyperaceae is considered to have been derived from axile or free central ancestry. This distinguishes it from the Gramineae where the so called basal ovule is derived from parietal ancestry.

Palynology in Relation to Taxonomy: The publication of *Pollen Morphology and Plant Taxonomy* by Erdtman in 1952 marked the beginning of a new phase. He made available pollen characters of all angiosperm families to taxonomists. Since then they are increasingly used in systematic work. Erdtman (1963) has given an excellent review of the systematic applications of palynology.

In addition to size and general shape, pollen characters such as number and position of the furrows, number and position of the apertures and the details of sculpturing of the exine are of taxonomic value.

Size and Shape: There is a considerable variation in the size of pollen in different plants. Several size classes, such as very small, small, medium, large, very large and gigantic have been recognised depending on the length of pollen.

The shape of pollen varies in different views. In polar view it may be circular, triangular or in other geometrical shapes, whereas in equatorial view it may be peroblate, prolate, spheroidal, subprolar, prolate or perprolate. The bilateral pollen are planoconvex, concavo-convex or biconvex in lateral view.

Furrows: Two basic types of pollen, monocolpate and tricolpate occur in the angiosperms. The former, which are characteristic of the monocotyledons and some dicotyledonous families such as woody Ranales, have a single furrow on one side of the pollen remote from the point of contact in the tetrad. The latter, which are characteristic of most of the dicotyledons, are provided with three meridionally placed furrows. Two other types of pollen, acolpate, without a furrow and pancolpate, with many furrows, are considered to have been derived from both the monocolpate and tricolpate types as a result of specialization.

Apertures: The form, number, distribution and position of apertures are important palynological criteria in assessing relationships and phylogeny of plants. There are two basic forms of aperture, the colpate or furrow form and the porate or circular form. An additional morphoform derived from them is the inaperturate form. The apertural morphoforms are divided in several subcategories on the basis of the nature of endoapertures.

The number of apertures varies from one to many. According to the position, the aperture may be proximal, distal, zonal or global. In terms of evolution the proximal position is most primitive and the global, most advanced. Distal and proximal positions are found in one aperturate forms and zonal and global in many aperturate forms. The distal position of aperture is characteristic of pterido-phytes, the proximal of gymnosperms and monocotyledons, and the zonal and global of dicotyledons.

Exine Sculpturing: The pollen in anemophilous group is smooth and thin, whereas the ones distributed by insects or birds are characterised by various types of sculpturing on the exine. The exine ornamentation may be of depression type or of excrescences type. The former includes many subtypes such as reticulate (lumina in the form of a net work), retipliate (incomplete fusion of columella), foreolate (circular and closely placed lumina), scrobiculate (circular and distantly placed lumina), fossulate (elongated lumina), striate (parallel lumina), regulate (anastomosing lumina) and areolate type (luminoid network surrounds islands of raised areas).

The various subtypes of excrescence type are: granulose (excrescences are minute granules), spinulose (excrescences are as spinules), gemmate (excrescences are in the form of rounded warts with constricted base), verrucate (excrescences are rounded but base is not constricted), tuberculate (tuberculer-like excrescences), spinose (long pointed excrescences), baculate (rod shaped excrescences) and clavate (club shaped excrescences).

Ultrafine Structure: In recent years finest details of surface reliefs of pollen have been obtained by transmission electron microscopy. Ultrafine structure of sporoderm has also been studied by means of ultra thin sections examined under electron microscope. Ultraviolet microscopy has proved useful in the study of fine details of the intine and cytoplasm. In Asteraceae, ultra structure of pollen has provided data of considerable taxonomic value.

Other Pollen Characters: Dufou (1961) has considered the systematic value of the nuclear condition of angiosperm pollen at the time of anthesis. Usually monocotyledons are 2-nucleate, polypetalous dicotyledons 2-nucleate, gamopetalous dicotyledons 3-nucleate and apetalous taxa 2-nucleate at the time of anthesis.

Pollen characters have helped in differentiation of various taxa (*e.g.*, families, genera and species), classification of newly discovered taxa and in phylogenetic considerations. Some striking examples are given below.

Family and Tribal Level: The families may be stenopalynous or eupalynous. The stenopalynous families are constant in their pollen characters. For example, Brassicaceae are characterised by 3-colpate and reticulate pollen, Ericaceae have pollen tetrads, Gyrostemonaceae are with thick walled pollen and Poaceae have pollen with a single pore surrounded by an operculum.

On the other hand, pollen may vary in size, aperture, exine ornamentation, etc. in eupalynous families and may be of considerable value in distinguishing lower taxa within the family. Verbenaceae, Betulaceae and Arecaceae are some examples of eupalynous families.

There are two families, Araceae and Lemnaceae in the order Arales of Hutchinson. The former is stenopalynous with 1-2-4-colpate, 3-porate or inaperturate pollen with Arecaceae exine sculpturing and the latter eupalynous with 1-porate and spinose pollen.

The tribe Bombaceae of the family Malvaceae has been recognised as a separate family Bombacaceae. The palynological studies also support this treatment. The exine in the Bombaceae is reticulate, whereas it is spinose in most of the Malvaceae.

The family Berberidaceae is variously circumscribed by the taxonomists. In recent treatments *Podophyllum* has been removed to a separate family Podophyllaceae. The grains remain united in *Podophyllum,* whereas they are free in the other Berberidaceae.

Pollen characters have also aided in more natural grouping of various genera in the Asteraceae. The sculpturing of exine in the Asteraceae is highly elaborate adapted to insect pollination. However, in anemophilous genera there are tendencies to simplification and loss of sculpturing.

Generic Level: The genera *Salix* and *Populus* (Salicaceae) can be distinguished on the basis of pollen characters. *Salix* has long and narrowed 3-furrowed pollen as compared to spherical pollen without distinct apertures in *Populus.*

The thickening of exine around the pores make a distinguishing character for different genera of the Betulaceae. In optical section it is knob-like in *Betula,* club-shaped in *Corylis,* unexpanded in *Caprinus,* and an arcus is present between adjacent pores in *Alnus.*

The two genera of the Phytolaccaceae, *Phytolacca* and *Rivinia* can be recognised on the basis of palynological characters. The pollen of *Phytolacca* is 3-zonocolpate, whereas that of *Rivinia* is pentocolpate.

Pollen characters have been extensively employed in classifying the genera of the Acanthaceae and the Primulaceae.

Pollen morphology has played an important role in the splitting of the genus *Polygonum* which contains several different types. Hedberg (1946) suggests division of the genus into seven genera: *Koenigia, Persicaria, Polygonum, Pleuropteropyrum, Bistoria, Tiniaria* and *Fagopyrum* which are distinct in their pollen types. This treatment is accepted by several recent taxonomists.

Species and Infraspecific Level: Pollen characters have proved helpful in distinguishing the species within a genus. Annual species of *Ranunculus* can be successfully identified with the help of pollen in the absence of fruits.

Species of *Anemone* can be distinguished on the basis of germinal aperture of pollen. It is 3-zonocolpate in *A. obtusiloba,* pentoporate in *A. alchemillaefolia,* pentocolpate in *A. rivularis* and spiraperturate in *A. fulgens.* Similarly, variations in the aperture and exine ornamentation may help in the identification of various species of *Bauhinia.*

Pollen size is helpful in distinguishing two species of *Malva, M. rotundifolia* (pollen 74-84 μ m) and *M. sylvestris* (pollen 105-126 μ m).

Sometimes pollen variations may help in interpretation of relationships of varieties in some cultivated plants such as *Hibiscus, Bougainvillea* and *Canna.*

Taxonomic Position of Disputed Taxa: The examination of the pollen sometimes helps in establishing correct position of disputed taxa. *Berenice arguta* was originally placed in the Saxifragaceae. Pollen examination, however, showed the presence of campanulaceous pollen and the examination of other features also indicated its correct position in the Campanulaceae. *Trisyngyne* is another interesting example. The male specimens of this taxon were placed in the Euphorbiaceae by Baillon. When the female specimens were found *Trisyngyne* was transferred to the Fagaceae. Examination of pollen showed that they are exactly similar to those of *Nothofagus* (Fagaceae).

Phylogenetic Considerations: Palynological studies have also helped in the elucidation of phylogenetic relationships. Contrary to most systems of classification, the pollen morphology does not support sharp demarcation between the dicotyledons and the monocotyledons as dicotyledonous pollen characters occur in some monocotyledons and vice versa.

Palynological studies suggest two distinct phylogenetic stocks in the dicotyledons, monocolpate and tricolpate represented by the Magnoliaceae and the Ranunculaceae respectively. The presence of monocolpate element in the monocotyledons indicates that they are more closely related to the magnolian stock. Furthermore, the monocotyledons and the magnolian dicots (both have monocolpate elements characteristics of the pre-angiospermous archegoniates) are considered more ancient palynologically than the ranalian dicots where monocolpate elements are completely absent and new apertural forms are present.

Kuprianova (1948), who studied the morphology of pollen in the monocotyledons considers that the Helobiae are not related to the other monocotyledons but are specialized Polycarpicae with ranalian affinities. She also pointed out that most monocotyledonous families could be considered to have evolved from Arecaceae or Liliaceae.

Embryology in Relation to Taxonomy: Embryology has also played a significant role in systematic considerations. Although some earlier botanists like Hofmeister and Strasburger indicated the possibility of utilizing embryological characters in taxonomy, it was Schnarf (1931) who brought to prominence the role of embryology in taxonomy. A large amount of work has subsequently been done in this direction where embryological characters are of value in classification, particularly of the larger taxonomic entities such as genera and tribes and in determining the relationship of families.

There are several embryological characters which have proved of taxonomic significance. Maheshwari (1950, 1963) has listed twelve features judged to be of special interest in taxonomic considerations in delimiting the larger plant groups :

(1) Number and arrangement of loculi in anther, structure and thickenings of endothecium and nature of tapetum.

(2) Mode of quadripartition of the microspore mother cells.

(3) Development and organization of the pollen grain.

(4) Development and structure of the ovule.

(5) Form and extent of the nucellus.

(6) Origin and extent of the sporogenous tissue in the ovule.

(7) Time of wall formation in megasporogenesis, arrangement of megaspores and development of the embryo sac.

(8) Form and organisation of the mature embryo sac.

(9) Fertilization-pathway of pollen tube and interval between pollination and fertilization.

(10) Type of endosperm and presence and absence of endosperm haustoria.

(11) Form and organization of the mature embryo.

(12) Certain abnormalities of development such as parthenogenesis, apogamy, adventive embryony and polyembryony.

There are certain families which are specially marked out by their embryological features that could identify these members. For example, Podostemonaceae are marked by the presence of pseudo-embryo sac, bisporic type of embryo sac, occurrence of pollen grains in pair, tenuinucellar bitegmic ovule where only outer integument forms the micropyle and the presence of prominent suspensor haustoria.

A characteristic embryological feature of the Onagraceae is the universal occurrence of the Oenothera type of embryo sac which is not found in any other family except an abnormality. The four nuclei formed after the two nuclear divisions remain together at the micropylar end of the embryo sac, the three organizing into an egg apparatus and the fourth functioning as a polar nucleus which is diploid.

In all taxa of Cyperaceae only one functional microspore is formed in each pollen mother cell instead of normal four microspores as in the vast majority of angiosperms. On the basis of this single feature it is possible to identify a member of the Cyperaceae. Similarly, the embryology of the Orchidaceae is characteristic in the absence of an endosperm.

The following are some noteworthy examples of the value of embryology in taxonomy:

Paeonia: There has been some controversy regarding the systematic position of *Paeonia*. In earlier systems of classification it has been included in the family Ranunculaceae. However, subsequently (eg. Hutchinson, 1959) it has been removed from Ranunculaceae and placed in a separate family, Paeoniaceae. Embryological evidence also supports the latter treatment. In contrast with other genera of the Ranunculaceae, in *Paeonia* the generative cell is large and elongated, the embryo sac is long and narrow, the seed coat is massive and the germination is hypogeal. The embryogeny of *Paeonia* is also unique and is not found in any other member of the Ranunculaceae.

Trapa: The genus *Trapa* was assigned to the family Onagraceae in Bentham and Hooker's system of classification. In Englerian system it was put in a separate family, a treatment followed by most of the recent systematists. This view is further confirmed on the basis of the embryological evidence. While the Onagraceae are characterized by the presence of Oenothera type of embryo sac, it is Polygonum type in *Trapa*. The endosperm is lacking in *Trapa* but a nuclear endosperm is present in the Onagraceae. The development of embryo is of Solanad type in *Trapa* and that of Onagrad type in the Onagraceae. The suspensor is well developed and haustorial in *Trapa* while it is short and inconspicuous in the Onagraceae. *Trapa* has very unequal cotyledons in contrast to equal cotyledons in the Onagraceae.

Butomus: While redefining the Butomaceae, Pichon (1946) suggested that all genera of the Butomaceae except *Butomus* should be transferred to the Alismataceae. Cronquist (1968, 1981) and Takhtajan (1969, 1980) also retained only *Butomus* in the Butomaceae and transferred other genera to the Limnocharitaceae. On embryological consideration also *Butomus* stands apart from the other members of the family which justifies its segregation. While *Butomus* has a Polygonum type of embryo sac, the other genera are characterized by an Allium type of embryo sac.

Peganum: The genus *Peganum* has been included in the Rutaceae by Bentham and Hooker (1862-1867) and in the

Zygophyllaceae by Engler and Prantl (1897). The course of embryo development of *Peganum* differs from that of the Rutaceae and the Zygophyllaceae. On the other hand, it is very close to that of the Linaceae. Thus the earlier placing of *Peganum* either in the Rutaceae or the Zygophyllaceae is not justified. Soueges (1953) suggests that it should be placed in a separate family, Peganaceae near Linaceae.

Parnassia: The systematic position of the genus *Parnassia* has been a matter of controversy. It has generally been included in the family Saxifragaceae. The embryogeny of *Parnassia* is far removed from the other members of the Saxifragaceae and hence it has been suggested that the genus should be placed in a separate family, Parnassiaceae.

Crassulaceae: The Crassulaceae resemble the Saxifragaceae in a number of embryological characters such as the presence of multilayered anther wall, crassinucellate bitegmic ovules, Polygonum type of embryo sac development, cellular endosperm and endosperm haustoria and Sedum variation of Caryophyllad type of embryo development. These justify placing of the Crassulaceae in the order Rosales close to the Saxifragaceae as has been envisaged by some earlier taxonomists.

Loranthaceae: Embryologically the two subfamilies of the Loranthaceae, Loranthoideae and Viscoideae are quite distinct. Besides floral structure, they differ in the mode of development of the embryo sac, endosperm and embryo. Furthermore, while in the Loranthoideae the viscid zone of the fruit is situated outside the vascular bundles of the floral tube, it is located inside the vascular bundles in the Viscoideae. On the basis of these differences it has been suggested that the two subfamilies be raised to the status of families.

Liliaceae: Embryological studies in the Liliaceae and their allies have aided in the taxonomy of this much debated group. Cave (1951) has suggested regrouping of genera on the basis of embryological characters.

The Liliaceae and Amaryllidaceae are considered as polyphyletic families, and certain groups of both these families are related to each other. Hutchinson (1959) has emphasized these resemblances and has united into the Agavaceae three tribes of the subfamily Dracaenoideae of the family Liliaceae, Yucceae, Nolineae and Dracaeneae (with the exception of two genera *Astelia* and *Milligania)* together with the subfamily Agavoideae from the Amaryllidaceae and *Phormium* of the tribe Hemerocallideae. He recognised six tribes in the Agavaceae:. Yucceae, Dracaeneae, Phormieae, Nolineae, Agaveae and Polyantheae. Comparative embryological study of Wunderlich (1950) concludes that Hutchinson's Agavaceae is not a uniform family and that the question whether these fairly heterogeneous groups should be united in one family remains open. She suggests that Hutchinson's Agavaceae should be separated into four groups: (1) Yucca-Agave, (2) Nolineae, (3) Dracaena-Sanseviera, and (4) Cordyline, and the genera *Phormium* and *Doryanthes* should be excluded.

Lemnaceae: The Lemnaceae are considered to have been derived either from the Araceae or from the Helobiales. Embryological study of S. C. Maheshwari (1954, 1956, 1958) clearly indicates that the Lemnaceae are evolved from the Araceae and they are not related to the Helobiales. The Lemnaceae resemble Araceae in the presence of bitegmic ovules with a micropylar cap of nucellus, cellular endosperm with a chalazal haustorium, irregular sequence of the divisions in the development of embryo and short and stocky suspensor and integumentary operculum in the seed. On the other hand, in the Helobiales the ovules do not have any micropylar cap, the endosperm is helobial or nuclear, the sequence of divisions in the development of embryo is very regular and a prominent basal cell in the embryo becomes hypertrophied.

Cyperaceae: The families Cyperaceae and Gramineae were considered to be closely associated by earlier systematists. Several morphological and anatomical features and the structure of the spikelets show that the two families are quite distinct. The embryological findings also do not corroborate

Cyperaceae-Gramineae alliance (Shah, 1967). In the Cyperaceae the microsporogenesis is of simultaneous type and the tetrad members do not separate (three degenerate and only one is functional), whereas in the Gramineae microsporogenesis is of successive type and the tetrad members separate. In the archesporium of the Cyperaceae the parietal cell is present while it is absent in the Gramineae.

The antipodals are ephemeral in the former, they form a complex or become coenocytic in the latter. The embryogeny of the Cyperaceae is of Onagrad type with Juncus variation and that of the Gramineae is variable. The testa and pericarp are distinctly free in the Cyperaceae and the former comprises both the integuments, whereas in the Gramineae testa and pericarp are fused and either both the integuments are obliterated or only the inner integument forms the testa.

On the other hand, embryological evidence shows a close relationship between the Cyperaceae and the Juncaceae. Although the Juncaceae produce pollen grains in tetrads, actually it is a monad also containing the three non-functional nuclei of the other three microspores of the tetrad. The pollen of the Cyperaceae is considered to be a reduced form of that of the Juncaceae. Further more, the two families resemble in the simultaneous cytokinesis, tenuinucellar ovules with obturator, Polygonum type of embryo sac with ephemeral antipodals, Onagrad type of embryogeny with Juncus variation and periclinal division of the cells of quadrat.

Cytology in Relation to Taxonomy: Cytology has made an outstanding contribution to taxonomy during the past few decades. Cytological characters such as chromosome number and morphology and chromosome behaviour and structure at meiosis are of considerable taxonomic value. Since the cytological data are directly derived from nucleus, the seat of the hereditary material, they may be used for understanding of the evolution and relationships of populations.

Chromosome Number: The chromosome number is usually constant in a species and this makes it an important taxonomic character. The lowest chromosome number in flowering plants is recorded in *Haplopappus gracilis* (2n=4) and the highest in *Poa litorosa* (2n=265). The chromosome numbers of species are included in many of the recent floras.

Sometimes the chromosome number is constant throughout the whole group, *e.g.*, *Quercus* and other members of the Fagaceae have the same basic number, n = 12. In such cases chromosome number is not of any help in distinguishing various taxa within the group.

In several cases a polyploid series is found in various members of a group. The basic number of chromosomes in *Salix* is 19 and the following species form a polyploid series : *S. viminalis* (diploid, 2n= 38), *S. atrocinerea* (tetraploid, 2n=76), *S. phylicifolia* (hexaploid, 2n=114) and *S. myrsinites* (octoploid, 2n=152). Another interesting example of polyploid series is found in the genus *Taraxacum* with 2n=16, 24, 32, 40 and 48. In *Rubus* (basic number 7), an apomictic genus, various apomictic biotypes are known with numbers of 14, 21, 28, 35,42 and 49. Thus chromosome numbers may provide a clue to reproductive irregularities as apomixis. Although the significance of polyploidy in taxonomy is debated, some cytologists are of the opinion that all taxa differing in chromosome number to be regarded as distinct species. Polyploidy is of wide occurrence in angiosperms and about half of the species sampled are polyploids.

When in a polyploid series a plant possesses an exact multiple of the basic number of chromosome, it is called *euploid*. On the other hand, when the chromosome numbers found within a group do not show simple numerical relationship to each other, it is said to be *aneuploid*. Sometimes an increase (through polyploidy followed by loss of chromosomes) or decrease (through unequal interchange between chromosomes) in basic number of chromosomes may arise. Individuals with 2n+1 are known as *trisomics* and those with 2n—1 as *monosomics*. In trisomics the extra chromosomes produce a certain amount of unbalance and thus limiting their taxonomic significance.

Normally diploid monosomics are inviable. Several disomic triploids have been recognised in *Taraxacum.*

Another type of supernumerary chromosomes found in plants are B-chromosomes. These chromosomes which are usually small and heterochromatic do not pair with normal A-chromosomes. They can, however, pair with each other. They reduce fertility or increase the vigour of plants and are restricted to certain families.

Chromosome Morphology: In addition to variation in number, chromosomes vary in form, size, volume, and in the amount of distribution of heterochromatin. These characteristics of karyotypes are taxonomically useful where the individual chromosomes are large enough for detailed microscopical observation.

The monocotyledons have usually larger chromosomes than the dicotyledons. In general, woody plants have smaller chromosomes than in their herbaceous relatives. Evolutionary trends in chromosome size have been worked out in some taxa.

The individual chromosomes of some taxa show marked differences in shape and size at mitotic metaphase. They are characterised by the relative length of the arms. Position of the centromere and presence of satellites mark further differences. Forms of chromosomes differ considerably and they may be symmetrical or asymmetrical. The former, known as V types, are with two equal arms and a median centromere. Of the latter, J types are with a long and a short arm and I types are rod shaped with a subterminal centromere. All intermediates between these three main types are found. In some taxa evolutionary trends can be worked out in these features. The Ranunculaceae are a very good example where such trends are found.

Accessory or B-chromosomes, which are usually small and heterochromatic, occur in different frequencies in some plant populations. At meiosis they do not pair with normal A-chromosomes. There is certain correlation between different frequencies of B-chrornosomes and some ecological factors.

Behaviour at Meiosis: A study of chromosomes behaviour at meiosis can provide some valuable information about the relationship of populations and species. The kind and degree of pairing show whether hybridisation has occurred, indicate structural differences in the parental chromosomes, and explain causes of sterility. The degree of chromosome homology in hybrids is an indication of the degree of relationship of the parental species.

Cytological Variation at Family Level: Cytological evidence is of considerable importance in interpreting a classification and establishing relationships. It may indicate direction of evolutionary change. Warburg (1938) has discussed taxonomy and relationships of the Geraniales on the basis of cytology.

Although taxonomic value of cytological data varies from group to group, they have provided logical basis for improved arrangement of tribes and genera in some families such as the Ranunculaceae, Brassicaceae and Poaceae.

In the Ranunculaceae chromosome number and chromosome morphology have provided basis for more natural arrangement of genera and tribes. Two major tribes of the Ranunculaceae, Helle-boreae and Anemoneae as recognised by Engler and Prantl have genera with base chromosome numbers of 7, 8 and 9 and the both have genera with large and small chromosome types.

The genera *Aquilegia* and *Isopyrum* (Helleboreae) and *Thalictrum* and *Anemonella* (Anemoneae) have base number 7 with small type chromosomes. These genera have thus been segregated in another tribe Thalictreae. Two other genera of the Helleboreae, *Coptis* and *Zanthorhiza* with very small chromosomes and the base number 9 have been removed to an additional tribe, Coptideae. Karyotypically *Nigella* and *Adonis* with base number 6 are distinct from other genera in family.

Paeonia, which has been removed to a separate family, is also distinct karyotypically having base number 5.

The sub-divisions of the family Brassicaceae as recognised by Schulz on traditional characters are also distinct karyotypically in having different base chromosome numbers as has been shown by the studies of Manton (1932).

The major sub-divisions of the family Gramineae as recognised currently are characterised by the number and size of the chromosomes.

The genera *Agave* and *Yucca* were placed in the separate families, Amaryllidaceae and Liliaceae respectively by some earlier workers have 5 long and 25 short chromosomes. Their karyotype similarity justify their inclusion in the family Agavaceae as has been done by Hutchinson.

Cytological Variation at Generic Level: There are several examples where chromosome numbers support generic status. In the tribe Boronieae of the Rutaceae different base numbers forming a more or less continuous series from 7 to 19 are characteristic of various genera.

The genus *Cistus* (Cistaceae) was previously included in *Helianthemum*. The former has base chromosome number 8 and the latter 9. This supports the recognition of *Cistus* as a separate genus.

Cicendia filiformis and *Microcala pusilla* (Gentianaceae) were previously placed in the same genus *Cicendia*. They have base chromosome number 13 and 10 respectively and this supports their generic separation.

Chromosome number is of little value if both aneuploidy and polyploidy occur in a genus, as in *Crepis*. However, if the chromosome number is constant within a genus, this is particularly useful in generic limitation. For example, in the genus *Tephrosia* (Fabaceae) all species are with 2n=22 except for *T. constricta* with 2n=16. This species has been separated as a distinct genus, *Sphinctospermum* and the chromosome number supports this separation.

Chromosome morphology has greater possibilities in generic limitations. A new genus and species, *Melanidion boreale* was described in Brassicaceae. Subsequently, the same species was described as *Acroschizocarpus kolianus*. On further observation it was found to be a species of an already established genus *Smelowskia*. The number, size-range and morphology of the chromosomes of what was described as *Melanidion* and *Acroschizocarpus* fit the pattern for *Smelowskia*.

The two genera of the Brassicaceae, *Physaria* and *Lesquerella* were recognised by many as a single genus. Cytological evidence, however, suggests that these two genera should remain separated.

Chromosome morphology has been effectively used in delimiting and in taxonomy of some genera such as *Nicotiana* and *Crepis* which are extensively investigated cytologically.

Cytological Variation at Specific and Infra-specific Level : A great deal of cytological variation occurs at specific and infra-specific level. The following are just a few examples to cite from the vast literature that has accumulated over the years.

On the basis of karyological studies, supplemented with other evidence, Fernandes (1951) has proposed a new classification of the genus *Narcissus* (Amaryllidaceae) and he has also discussed phylogenetic relationships of the various species of the genus.

Monotropa hypopitys (Monotropaceae) was formerly treated a single species with two varieties, var. *hirsuta* and var. *glabra*. On cytological examination the former was found to be a hexaploid, with 2n = 48, and the latter a diploid, with 2n = 16. The hexaploid was retained as the species *M. hypopitys* and the var. *glabra* was raised to the specific rank as *M. hypophegea*. Brandt (1961) recognised two cytological races in *Veronica prostrata* (Scrophula-riaceae) and suggested that they should be treated as subsp. *prostrata* (n = 8 and with small pale blue flowers) and subsp. *scheererl* (n=16 and with larger dark blue flowers).

Chemotaxonomy

The application of chemistry to systematics is chemotaxonomy or chemical taxonomy. Chemical characters of plants have long been of practical value. For example, the smells arising from the crushed foliage (due to the presence of characteristic essential oils) of the Apiaceae and Lamiaceae are quicker means of identifying their members, and the two subfamilies of the Asteraceae, the Tubiflorae and Liguliflorae are distinguished on the basis of presence and absence of latex respectively.

In recent years chemotaxonomy has made a significant impact on taxonomic and evolutionary studies. Chemical characters of plants can often find as wide application in classification as do characters from gross morphology. They have provided useful information for the understanding of evolutionary relationships of various taxa and have also helped in the solution of many taxonomic problems.

The occurrence and distribution of the various types of chemical substances present in plants form the taxonomic evidence. However, all kinds of chemical substances do not reveal information useful to the taxonomist. Distribution of secondary compounds of low molecular weight such as nonprotein free amino acids, phenolics and betalins, alkaloids, terpenoids and steroids provide valuable clues to the systematist. Chemical evidence is applied by the taxonomist at all levels of the hierarchy of classification from varietal rank to the divisions of plants. Chemical data are also a reliable guide to phylogenetic relationships of various taxa.

Some of the important sources of taxonomic evidence are discussed briefly in the following pages:

Amino Acids: There are two groups of amino acids—protein amino acids and non-protein amino acids. The former are the building units of proteins and they are released on hydrolysis of proteins. The latter, which are so far not found in combination in proteins, are more numerous and are estimated to be about 300 in number. Discontinuous distribution and less susceptibility to rapid change increase the taxonomic value of non-protein amino acids.

Distribution of non-protein amino acids in Fabaceae has been of considerable interest to the taxonomist. They are present in unusually high concentrations in the seeds. Bell (1962) has recognised seven infrageneric groups in the genus *Lathyrus*. Each group is characterized by a different amino acid or group of amino acids. Similarly, four infrageneric groups have been recognised in the genus *Vicia* on the basis of distribution of amino acids.

The amino acid content of seeds in the section Gummiferae of the genus *Acacia* is so different from that in the other sections that the species of this section can be recognised on the basis of this single character alone. Free amino acid, lathyrine is so far known only in the genus *Lathyrus* and this supports the present circumscription of the genus. Azetidine-2-carboxylic acid is extremely restricted in distribution. It is characteristic of many genera of the Liliaceae, Amaryllidaceae and Agavaceae which are related families.

Phenolics: Phenolic compounds are present in plants in considerable number, diversity and quantities. Their quick extraction and easy separation by paper chromatography make them popular from taxonomic view point.

Flavinoids are the largest group of naturally occurring phenols and they are most often studied in chemical taxonomy. They are mostly found in the vacuole of the plant cell. Various types of flavinoids include flavones, flavanones, isoflavones and isoflavonoids, flavonols, anthocynidins, chalcones and aurones, and biflavonyls. Flavinoids from vegetative parts and seeds provide more reliable taxonomic evidence than flower pigments which are too variable.

The presence of leuco-anthocyanin is correlated with woody habit. They were detected in 60% of the woody families examined, whereas they were observed in only 15% herbaceous families. The presence of flavonols and methoxycinnamic acids is also correlated with the woody and herbaceous habit.

The discontinuous distribution of phenolics provides valuable taxonomic evidence. For example, ellagic acid is present only in the tribe Kerrieae of the subfamily Rosoideae of the family Rosaceae, whereas it is absent in other tribes of the subfamily. Isoflavone iridin is known only in the Pogonivis section of *Iris. Iris flavissima* which was originally placed in section Pogonivis does not contain iridin and on the other hand resembles species of Regelina section in phenolic characters. This favours transfer of *Iris flavissima* to Regelina section.

The South American species of *Eucryphia* (Eucryphiaceae) are very similar to the Australian species morphologically but they can be easily distinguished on the basis of the flavonoid glycosides.

Betalins: There is a new class of pigments which are not flavinoids but are functionally equivalent to phenolics. Due to the presence of nitrogen they are excluded from the' general definition of phenolics. These red and yellow pigments (betacyanins and betaxanthins) are confined to ten families of angiosperms (Chenopodiaceae, Portulacaceae, Aizoaceae, Cactaceae, Nyctaginaceae, Phytolaccaceae, Stegnospermaceae, Basellaceae, Amaranthaceae and Didieraceae) which have been included in a single order, the Centrospermae. Betalin-containing plants do not contain anthocyanins which are the normal pigments found in other angiosperm families. The two groups are unrelated both chemically and biosynthetically.

The distribution of betalin constitutes a notable chemotaxonomic contribution. Of the ten families which contain betalin, the systematic position of the Cactaceae has been disputed in the past. They are often placed in an order of their own, the Cactales or Opuntiales. The presence of betalin in the Cactaceae establishes their position in the Centrospermae. The chemical evidence suggests limiting of the Centrospermae to only those families which contain betalin, while the families Caryophyllaceae, Illecebraceae and Molluginaceae (which lack betalin and contain anthocyanins) should be placed in a separate order, the Caryophyllales.

Alkaloids: Alkaloids are a heterogeneous group of organic nitrogen containing bases, often with a heterocyclic ring. They are not essential to the growth of plant and exert marked physiological effects on the nervous system of animals. True alkaloids have a nitrogen-containing heterocyclic nucleus derived from a biogenetic amine and they can be related structurally to parent bases such as pyridine, piperidine, isoquinoline and tropane. Alkaloids are distributed throughout the plant tissues and are present in the vacuoles in the form of salts. There are over 5000 alkaloids reported from angiosperms mostly from the dicotyledons.

Families like Berberidaceae, Fabaceae, Ranunculaceae and Solanaceae are specially rich in alkaloidal species. Since these families are sufficiently different, the presence of alkaloid does not imply taxonomic relationship at family level. Alkaloid content can be considered as a source of taxonomic evidence as chemical or biogenetic group of species of a particular taxon is the same. For instance, the members of the Papaveraceae synthesize isoquinoline alkaloids, those of the Fabaceae lupin alkaloids and those of the Solanaceae tropane derivatives.

Sometimes alkaloids have a very narrow distribution, for example morphine is restricted to *Papaver somniferum*, coniine to a few Apiaceae and strichnine to a few species of *Strychnos*. Distribution of alkaloid has proved useful in the taxonomy of the Fabaceae. Three genera of Fabaceae, *Genista*, *Ammodendron* and *Adenocarpus* contain ammodendrine-hystrine alkaloids. Of these, *Genista* and *Adenocarpus* were included in the Genisteae, whereas *Ammodendron* was placed in the Sophoreae, a tribe characterized by the presence of matrine alkaloids. This suggests the transfer of *Ammodendron* to the tribe Genisteae.

The tribes Sophoreae, Genisteae and Podalyricae of the subfamily Lotoideae of the Fabaceae are characterised by the presence of lupin alkaloids and this suggests that these tribes may have originated from a common ancestral stock. The accumulation of isoquinoline alkaloids in the families Fumariaceae and Papaveraceae indicates very close relationship between the two families.

Terpenoids and Steroids: Terpenoids and steroids are a diverse group of compounds which are variously treated as alkaloids, glycosides or plant pigments. They are mostly polymerized isoprene residues. These compounds have very clear cut functions.

The carotenoids, a prominent group of terpenoids may prove of some taxonomic usefulness. Petal carotenoids may be helpful in the taxonomy of the tribe Genisteae of the Fabaceae and of the family Asteraceae.

Bate-Smith and Swain (1966) commented on the possible value to taxonomy of the iridoids (monoterpenoid cyclopentanoid lactones). Of these, arsperuloside is found in the Rubiaceae and acubin in the Cornaceae and the Scrophulariaceae, Orobanchaceae and some closely related families. In the 12th edition of Engler's Syllabus *Buddlela* has been transferred from the Loganiaceae to the Buddleiaceae (with a position near Scrophulariaceae) on the basis of the presence of acubin. On morphological evidence, removal of Garryaceae from apetalous orders to the vicinity of the Cornaceae is suggested in Engler's syllabus and this treatment receives support from chemical evidence as both Garryaceae and Cornaceae produce acubin.

On the basis of lactone distribution, the removal of *Ambrosia* and related genera from the tribe Heliantheae is suggested. These genera should be recognised as a separate tribe. The distribution of the steroid alkaloids in the tribe Veratreae of the Liliaceae supports the classification of genera as has been done on morphological criteria. The presence of triterpenoid glycosides cucurbitins in the Cucurbitaceae is considered of taxonomic significance.

Other Chemical Substances: Some directly visible chemical substances such as starch grains and raphides also provide taxonomically valuable characters. For example, form of starch grains has been used as an aid to classification of the Poaceae along with other characters and the presence and absence of raphides has proved a useful character in the more natural classification of the Rubiaceae. Forms of the crystals of calcium oxalate found in the ovary wall of Asteraceae may be of considerable taxonomic value.

Serology and Taxonomy: Immune serum (antiserum) produces antibodies as a result of introduction of foreign cells or particles (antigens). Serology is the study of the origin and properties of antisera. Proteins, which carry useful taxonomic information and are easy to handle, are the most widely used antigens in serotaxonomy. Storage proteins are most amenable for taxonomic studies followed by pollen proteins. They are injected in emulsion form into test animals such as rabbit. The amount and nature of the precipitate formed by antigen-antibody complex is of potential taxonomic value.

Serotaxonomy has developed in Germany. This country has been active centre since the beginning of this century and numerous papers have appeared on plant relationships. Mez and Ziengenspeck (1926), on the basis of results of serological tests published the 'Stammbaum', a family tree of plant relationships. On the basis of protein (of mature seeds) comparison by serological methods, Jensen (1968) discussed the relationships and classification of 20 genera of the Ranunculaceae.

Close relationship of *Delphinium* and *Aconitum* is supported by serological evidence. *Hydrastatis,* which is sometime classified with the Berberidaceae has more close serological similarity to the Ranunculaceae than the Berberidaceae. *Trollis,* which is regarded as a link between follicular and achenial fruit bearing genera in the family, resembles closely with *Adonis.*

Serological methods have also proved useful in the taxonomy of legumes and potato, and in the classification of *Bromus.*

Binomial System

The binomial system of nomenclature was employed by Linnaeus in the first edition of his *Species Plantarum* published in 1753, although this system was also used to some extent by

Gaspard Bauhin as early as 1623 in his *Pinax*. According to the binomial system of nomenclature the name of a plant consists of two Latin or latinised words, the first (generic epithet) representing the genus and the second (specific epithet) the species. For example, the botanical name of mango is *Mangifera indica*. The first word *(Mangifera)* designates the genus of the plant and the second word *(indica)* a particular species of that genus.

Global Norms

The foundations of International code of botanical nomenclature are found in Linnaeus' *Philosophia Botanica* published in 1751 wherein he proposed certain principles of nomenclature. Another significant work on plant nomenclature was Augustin de Candolle's *Theorie elemaintaire de la botanique* (1813) which gives detailed rules on plant nomenclature. However, the first organised efforts to develop a precise and simple system of nomenclature to be used by botanists in all countries were made at the First International Botanical Congress held in 1867 in Paris.

At this Congress the Lois de la nomenclature botanique (Laws of botanical nomenclature) proposed by Alphonse de Candolle (son of Augustin de Candolle) were adopted with some modifications as guide for nomenclature in the vegetable kingdom. These rules are known as de Candolle rules or Paris Code of 1867.

The subsequent International Botanical Congresses made significant contribution with regard to plant nomenclature but it was only at the Cambridge in 1930 that "For the first time in botanical history, a code of nomenclature came into being that was international in function as well as in name" (Lawrence, 1951). The rules of nomenclature adopted at the Cambridge Congress were subjected to some changes and refinement from time to time and the current *International Code of Botanical Nomenclature* as appeared in 1978 was adopted by the Twelfth International Botanical Congress held in Leningrad (U.S.S.R.

in August 1975. The following information is based on this edition of the Code.

The code is divided into three parts—Principles, Rules and Recommendations.

The *Principles* form the basis of the system of botanical nomenclature. The six principles given in this section are:

(1) Botanical nomenclature is independent of zoological nomenclature.

(2) The application of names of taxonomic groups (taxa) is determined by means of nomenclature types.

(3) The nomenclature of taxonomic groups is based upon priority of publication.

(4) Each taxonomic group can bear only one correct name, the earliest that is in accordance with the Rules, except in specific cases.

(5) Scientific names of taxonomic groups are treated as Latin regardless of their derivation.

(6) The Rules of nomenclature are retroactive unless expressly limited.

The *Rules* give detailed prescriptions on all the points connected with the naming of plants. The names contrary to a rule can not be maintained.

The *Recommendations* are often practical application of rules. Their object is to bring about greater uniformity and clearness, especially in future nomenclature.

The code has three appendices:

Appendix I, deals with the names of hybrids.

Appendix II, Nomina Familiarum Conservanda, includes names of families which are conserved.

Appendix III, Nomina Generica Conservanda Et Rejcienda, lists the names of genera which are conserved against the principle of priority because of their long use.

Some Important Rules

1. *Ranks of Taxa:* According to Article 1 of the Code the word *taxa* (singular: *taxon)* signify 'taxonomic groups of any rank'. The rank of species is basic and the relative order of the ranks of taxa are: species, genus, family, order, class, division and vegetable kingdom. Thus each species is included in a genus, each genus in a family and so on.
2. *Typification:* According to Article 7, "The application of names of taxa of the rank of family or below is determined by means of nomenclatural types." The nomenclatural type is that element with which the name is permanently associated. The type of a species or infraspecific taxon is an individual specimen, however, for small herbaceous plants the type consists of more than one individual mounted on a sheet. If a specimen can not be preserved, the type may be a description or figure. The type of a genus is a species, that of a family a genus.

The following terms are used in the nomenclature of types :

(a) *Holotype:* It is the one specimen or other element designated by the author as the nomenclatural type.

(b) *Isotype:* It is a duplicate of holotype. For example, if several herbaceous plants (mounted on separate sheets) or several branches of a tree are collected at the same time, one is designated as type and the others become isotypes.

(c) *Lectotype:* It is a specimen or other element selected from the original material to serve as a nomenclatural type when no holotype was designated at the time of publication or as long as it is missing.

(d) *Syntype:* It is any one of the two specimens cited by the author when no holotype was designated or more than one specimens are designated as type.

(e) *Neotype:* It is a specimen or other element selected to serve as nomenclatural type as long as all of the material on which the description of the new species was based is missing.

3. *Priority:* Each taxon can bear only one correct name and the correct name is the earliest legitimate one except in case of limitation of priority by conservation. The principle of priority does not apply to names of taxa above the rank of family.

The date of valid publication of names of plants of different groups of Spermatophyta is treated as beginning 1 May, 1753, the date of publication of the first edition of Linnaeus' *Species Plantarum.*

4. *Names of Families:* The name of a family is a plural adjective and is formed by adding the suffix— *aceae* to the stem of a legitimate name of an included genus, *e.g.,* Rosaceae (from *Rosa)* and Cucurbitaceae (from *Cucurbita).*

Special exception is made for the following eight families. These names because of long usage are treated as validly published. For these families alternative names are also permitted ending in—*aceae.*

Palmae	(Arecaceae)
Gramineae	(Poaceae)
Cruciferae	(Brassicaceae)
Leguminosae	(Fabaceae)
Guttiferae	(Clusiaceae)
Umbelliferae	(Apiaceae)
Labiatae	(Lamiaceae)
Compositae	(Asteraceae)

5. *Name of Species:* The name of species is a binary combination consisting of the name of the genus followed by the specific epithet. If an epithet consists of two or more words, these are to be united or hyphened (*e.g., Hibiscus rosa-sinensis).*

The name of species may be taken from any source and may even be composed arbitrarily. However, the specific epithet should not be exactly the generic name, *e.g.*, names like *Linaria linaria* and *Phragmites phragmites*, where specific epithet exactly repeats the generic name are known as tautonyms. Tautonyms are treated as illegitimate in botanical nomenclature.

6. *Names of Infraspecific Taxa:* An infraspecific taxon (taxon below the rank of the species) when containing the nomenclatural type of next higher taxon can not be indicated by epithets such as *typicus, originalis*, etc. Article 26 requires that in such cases the specific epithet should be repeated unaltered in infraspecific taxon. For example, the combination *Lobelia spicata* var. *originalis* McVaugh which includes the type of *Lobelia spicata* should be named as *Lobelia spicata* Lam. var. *spicata.*

7. *Names of Plants in Cultivation:* Plants brought from the wild into cultivation retain original names. For example, forms of *Chrysanthemum parthenium* brought into cultivation are not to be renamed *Matricaria eximia.*

8. *Conditions of Effective and Valid Publication:* According to Article 29 of the code publication of new names and descriptions are effective when the printed matter is distributed to the general public or to atleast ten botanical institutions with libraries accessible to botanists generally. The date of effective publication is the date on which the printed matter became available.

   The following are important conditions attached for valid publication of names of new taxa:

   (1) Publication must be effective as indicated above.

   (2) It should be accompanied by a description or diagnosis or a reference to a previously and effectively published description or diagnosis.

   (3) It must be accompanied by a Latin description or diagnosis or by a reference to a previously published Latin description or diagnosis.

(4) A nomenclatural type is to be indicated for a new taxon of the rank of family or below published on or after 1 January, 1958.

9. *Retention of Names of Taxa Which are Divided* . When a genus is divided into two or more genera, the original generic name must be retained for the genus including the type species.

Similarly, when a species is divided into two or more species, the original specific name must be retained for the species containing the type. For example, *Lychnis dioica* L. was divided by Miller into two species which were named as *L. dioica* L. emend Mill and *L. alba* Mill.

The same rule also applies to infraspecific taxa.

10. *Changes in Names of Taxa:* Changes in the names of taxa may be required (a) by transference of the taxon, or (b) by its union with another taxon of the same rank, or (c) by a change of its rank.

(a) *Retention of Names of Taxa on Transference:* When a subdivision of a genus is transferred to another genus without change of rank, its original name (if legitimate) must be retained. For example, when *Saponaria* sect. *Vaccaria* DC. was transferred to *Gypsophila* it became *Gypsophila* sect. *Vaccaria* (DC.) Godr.

When a species is transferred to another genus without change of rank, the original name (if legitimate) must be retained. For example, *Hydrocotyle asiatica* L. when transferred to the genus *Centella* must be called *Centella asiatica* (L) Urban.

The same rule also applies to infraspecific taxa.

(b) *Choice of Names When Taxa of the Same Rank are United:* When two or more taxa of the same rank are united into one, the oldest legitimate name is retained for the combined taxon. If the names of the taxa are of the same date, the author who first unites them has the right to choose any one of the names, and his choice must be followed. For example, K. Schumann united the three genera *Sloanea* L. (1753), *Echinocarpus* Blume (1825) and *Phoenicosperma* Miq. (1865) and he has rightly adopted the name *Sloanea* L. for the resulting genus which is the oldest of the three generic names.

Robert Brown united *Waltheria americana* L. (Sp. Pl. 673, 1753) and *W. indica* L. (Sp. Pl 673, 17 53) and he adopted the name *W. indica* for the combined species which is to be retained.

(c) *Choice of Names When the Rank of a Taxon is Changed :* When the rank of a genus or a taxon below the rank of genus is changed, the correct name is the earliest legitimate one available in the new rank. For example, *Magnolia virginiana var. foetida* L. (1753) when raised to specific rank, its correct name should be *Magnolia grandiflora* L. (1882), not *M.foetida* (L.) Sargent (1889).

7

PROTEINS AND FATS

FAT METABOLISM

Fats and other lipids are important constituents of plants. Many seeds store large amounts of fats, which are utilised for the growth of the seedling during seed germination and early plant life. The fats are hydrolysed and the fatty acids and glycerol generated act as source of energy and carbon. The fats are highly concentrated stores of metabolic energy because they are reduced and anhydrous. For example, while complete oxidation of 1.0 g of fatty acid produces 9.0 kCal of energy, the same amount of carbohydrate produces only about 4.0 kCal. Lipids are also important constituents Of biological membranes as phospholipids and glycolipids. Many fatty acids derivatives serve as hormones and intracellular messenger.

Distribution of Fats

Fats are usually concentrated in the storage tissues of the plant, such as endosperm and cotyledons of the seed. Seeds of conifers, oil seeds and nuts are specially rich in fat. Some fruits are also rich in fat but very little fat is present in leaves, stems and roots. Fat is always stored in specialized bodies within the cells. These bodies have been called *lipid bodies, sphaerosomes* or *oleosomes.* These bodies are about 1 μm in diameter are hundreds of such bodies can be seen inside a single cell.

General Structure: Chemically, a neutral fat (true fat) is a triacylglycerol; where three fatty acid molecules are linked through ester linkage to a glycerol molecule. Therefore, neutral fats are also called as triglycerides.

$$H_3C-(CH_2)_7-CH=CH-(CH_2)_7-\overset{O}{\overset{\|}{C}}-O-\underset{H_2C-O-\underset{\underset{O}{\|}}{C}-(CH_2)_{16}-CH_3}{\overset{H_2C-O-\overset{O}{\overset{\|}{C}}-(CH_2)_{14}-CH_3}{C-H}}$$

A. triacyl glycerol

$$R_2-\overset{O}{\overset{\|}{C}}-O-\underset{H_2-C-O-\overset{O}{\overset{\|}{C}}-R_3}{\overset{H_2-C-O-\overset{O}{\overset{\|}{C}}-R_1}{C-H}}$$

B. General formula of a triacyl glycerol

The physical properties of the fat are determined by the kind of fatty acids they contain. Usually the three fatty acids are different, although sometimes two of them are identical. These acids almost always contain even number of carbon atoms usually 16 or 18 and they can be saturated unsaturated. When fat is to be used as a source of energy and carbon, it is first hydrolysed to fatty acids and glycerol. Fatty acid and glycerol are then oxidized to produce energy and other carbon compounds.

Hydrolysis of Fat

The initial event in the utilization of fat is its hydrolysis by the enzyme lipase;

$$\underset{\text{Fat}}{R_2-\overset{O}{\overset{\|}{C}}-O-\underset{CH_2O-O-\overset{O}{\overset{\|}{C}}-R_3}{\overset{CH_2O-O-\overset{O}{\overset{\|}{C}}-R_1}{CH}}} \xrightarrow{\text{Lipase}} \underset{\text{Glycerol}}{HO-\underset{CH_2OH}{\overset{CH_2OH}{CH}}} + \underset{\text{Fatty acids}}{\begin{matrix} R_1COOH \\ R_2COOH \\ R_3COOH \end{matrix}} + 3H^+$$

The presence of lipase has been reported in almost all kinds of organisms. In the seeds of oil plants such as castor bean, maize, cotton, mustard and rape the enzyme is associated with the membranes of lipid bodies which contain fats. The enzyme has been isolated from maize scutellum and purified to apparent homogeneity by Yon-Hui Lin and A.H.C. Huang (1984). It has a m.w. of about 270 kD and pH optimum of 7.5. The lipase activity is absent in ungerminated seeds and it appreciably increases during early growth of the seedling.

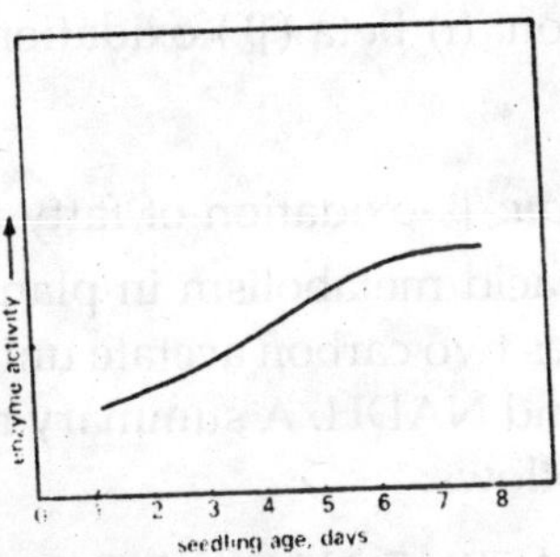

Changes in lipase activity during seed germination and early seedling growth.

Metabolism of Glycerol: Glycerol formed by the action of lipase on fats is phosphorylated and oxidised to dihydroxyacetone phosphate, which in turn is isomerized to glyceraldehyde 3-phosphate. This can be converted to pyruvate through glycolytic pathway;

(i)

$$\underset{\text{Glycerol}}{\begin{array}{l} CH_2OH \\ | \\ CHOH \\ | \\ CH_2OH \end{array}} \xrightarrow[\text{ATP} \quad \text{ADP}]{\text{Glycerol kinase}} \underset{\text{Glycerol 3-phosphate}}{\begin{array}{l} CH_2OH \\ | \\ CHOH \\ | \\ CH_2O(P) \end{array}}$$

(ii)

$$\underset{\text{Glycerol 3-phosphate}}{\begin{array}{l} CH_2OH \\ | \\ CHOH \\ | \\ CH_2O(P) \end{array}} \xrightarrow[\text{NAD}^+ \quad \text{NADH}]{\text{Glycerol phosphate dehydrogenase}} \underset{\text{Dihydroxy acetone phosphate (DHAP)}}{\begin{array}{l} CH_2OH \\ | \\ C{=}O \\ | \\ CH_2O(P) \end{array}}$$

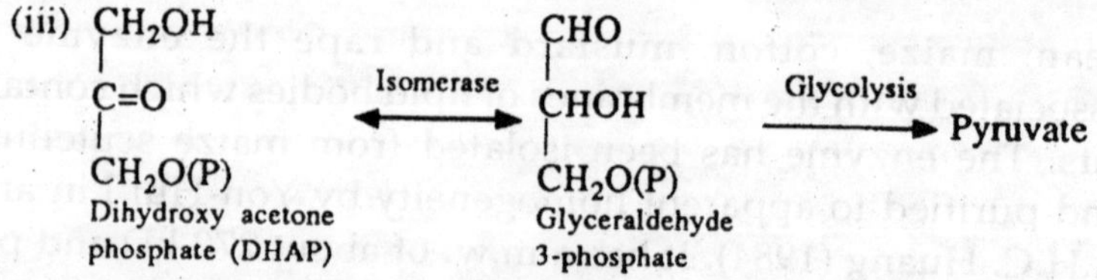

Oxidation of Fatty Acids: Fatty acids can be oxidised to CO_2 and H_2O, accompanied by the production of large amounts of ATP, NADH and $FADH_2$. There are two mechanisms of fatty acid oxidation: (i) Beta (β) oxidation and (ii) Alpha (α) Oxidation.

Beta Oxidation: β-oxidation of fatty acids is the primary pathway of fatty acid metabolism in plants and animals. The oxidation produces two carbon acetate units as acetyl CoA and releases $FADH_2$ and NADH. A summary reaction of palmitate oxidation is as follows:

$$\text{Palmitate} + \text{ATP} + 7\,\text{NAD}^+ + 7\,\text{FAD} + 7\text{H}_2\text{O} + 8\text{CoASH}$$

$$\rightarrow 8\text{Acetyl CoA} + \text{AMP} + \text{pyrophosphate} + 7\text{NADH} + 7\text{H}^+ + 7\text{FADH}_2$$

The removal of 2 carbon fragment, acetyl CoA, takes place in a sequence; at the end of each series of reactions, one acetyl CoA is removed. The reactions are as follows:

(1) First, the fatty acid is activated by linking to coenzyme A; energy is derived from ATP,

$$R\text{—}CH_2\text{—}CH_2\text{—}COOH + ATP + CoA\text{—}SH \xrightleftharpoons{\text{Acyl CoA synthetase}} R\text{—}CH_2\text{—}CH_2\text{—}\overset{\overset{\displaystyle O}{\|}}{C}\text{—}S\text{—}CoA + AMP + Pi$$

This activation reaction takes place on outer mitochondrial membrane. Then the activated fatty acid is carried across the inner mitochondrial membrane by carnitine. Further oxidative reactions take place inside the mitochondria.

(2) The acyl CoA is degraded by an acyl CoA dehydrogenase to give rise to an enoyl CoA;

$$R{-}CH_2{-}CH_2{-}\overset{O}{\overset{\|}{C}}{-}S{-}CoA \underset{FAD \quad FADH_2}{\xrightleftharpoons{\text{Acyl CoA dehydrogenase}}} R{-}CH_2{-}\underset{\underset{H}{|}}{C}={\overset{\overset{H}{|}}{C}}{-}\overset{\overset{O}{\|}}{C}{-}S{-}CoA$$

Enoyl CoA

In this reaction FAD accepts electrons and is reduced to $FADH_2$

(3) In the next step, the enoyl CoA is hydrated by enoyl CoA hydratase and hydroxy acyl CoA is produced,

$$R{-}CH_2{-}\underset{\underset{H}{|}}{C}={\overset{\overset{H}{|}}{C}}{-}\overset{\overset{O}{\|}}{C}{-}S{-}CoA \underset{H_2O}{\xrightleftharpoons{\text{Enoyl CoA dehydrogenase}}} R{-}CH_2{-}\underset{\underset{H}{|}}{\overset{\overset{\overset{H}{|}}{O}}{\overset{|}{C}}}{-}\underset{\underset{H}{|}}{\overset{\overset{H}{|}}{C}}{-}\overset{\overset{O}{\|}}{C}{-}S{-}CoA$$

L-hydroxy acyl CoA

(4) Next step is the oxidation (second time) reaction. The hydroxyl group is converted to keto group by the enzyme L-3-hydroxy acyl coA dehydrogenase and keto acyl CoA is produced. NADH is the coenzyme produced in this reaction,

$$\text{L-hydroxy acyl CoA} \xrightarrow[\text{NAD} \quad \text{NADH}]{\text{Hydroxy acyl CoA dehydrogenase}} R{-}CH_2{-}\overset{\overset{O}{\|}}{C}{-}CH_2{-}\overset{\overset{O}{\|}}{C}{-}S{-}CoA$$

Keto acyl CoA

(5) The final step in beta oxidation of fatty acids is the cleavage of keto acyl CoA by the thiol group of a

second molecule of CoA, which yields acetyl CoA and a fatty acid CoA shortened by two carbon atoms. This reaction is catalysed by ketothiolase,

$$\text{Keto acyl CoA} + \text{HS—CoA} \xrightleftharpoons{\text{Ketothiolase}} \underset{\text{Acetyl CoA}}{CH_3\text{—}\overset{O}{\overset{\|}{C}}\text{—S—CoA}} + \underset{\text{Fatty acyl CoA}}{R\text{—}CH_2\text{—}\overset{O}{\overset{\|}{C}}\text{—S—CoA}}$$

The shortened fatty acyl CoA then undergoes another cycle of oxidation involving all the steps (2) to (5). In each cycle, an acyl CoA is produced which is short by two carbon atoms and, one each of $FADH_2$, NADH and acetyl CoA is produced. The general reaction can be represented as follows;

$$\text{C(n)acyl CoA} + \text{FAD} + \text{NAD}^+ + H_2O + \text{CoA}$$

$$\rightarrow \text{C(n - 2)acyl CoA} + FADH_2 + \text{NADH} + \text{acetyl CoA} + H^+$$

The degradation of palmitoyl CoA (C16acyl CoA) requires seven reaction cycles. In the seventh reaction, the four carbon keto acyl CoA is cleaved to two molecules of acetyl CoA. The energy yield in terms of ATP in complete oxidation of palmitic acid molecule is as follows;

(1) $FADH_2$ in seven turns of oxidation 7 $FADH_2$, each $FADH_2$ produces two ATP on oxidation. Thus total $FADH_2$ is equivalent to 7 × 2 = 14 ATP

(2) NADH in seven turns of oxidation 7 NADH, each NADH produces three ATP on oxidation. Thus total NADH is equivalent to 7 × 3 = 21 ATP

(3) Acetyl CoA in seven turns of cycle 8 acetyl CoA, oxidation of each acetyl CoA through TCA cycle produces 12 ATP. Thus total acetyl CoA is equivalent to 12 x 8 = 96 ATP

(4) Total ATP produced 14 + 21 + 96 = 131

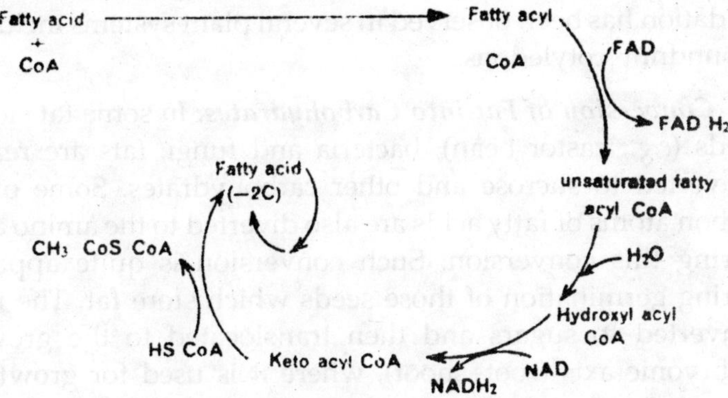

β *-oxidation cycle.*

(5) Total ATP consumed in reaction (in step 1) is two, as two high energy phosphate bonds are consumed in the activation of palmitate, in which ATP is split into AMP and 2 Pi.

(6) Thus, net gain in complete oxidation of one molecule of palmitate is 131-2= 129 ATP.

Alpha Oxidation: It is an alternate mechanism of oxidation of fatty acids, in which the acid is oxidized to produce another acid with one carbon atom less. This is because of elimination of CO_2. The process results in the production of fatty acids with odd number of carbon atoms.

The first step in alpha oxidation is decarboxylation of fatty acid. The product is an aldehyde with one carbon less than the fatty acid, and CO_2. The reaction is catalysed by a peroxidase type of enzyme;

$$R(CH_2)_nCH_2\text{--}COOH \xrightarrow[+O]{\text{Fatty acid peroxidase}} R(CH_2)_nCHO + CO_2$$

This aldehyde is then oxidized to the acid by dehydrogenase

$$R(CH_2)_nCHO \xrightarrow[NAD^+ \quad +O \quad NADH]{\text{Aldehyde dehydrogenase}} R(CH_2)_nCOOH$$

The new acid becomes the substrate for fatty acid peroxidase and takes another turn in alpha oxidation spiral. This kind of oxidation has been observed in several plant systems including groundnut cotyledons.

Conversion of Fat into Carbohydrates: In some fat storing seeds (*e.g.*, castor bean), bacteria and fungi, fats are readily converted to sucrose and other carbohydrates. Some of the carbon atoms of fatty acids are also diverted to the amino acids during this conversion. Such conversion is quite apparent during germination of those seeds which store fat. The fat is converted to sugars and then translocated to the growing embryonic axis (root+shoot), where it is used for growth.

The conversion takes place through *glyoxylate cycle,* as demonstrated by H.L. Kornberg and Harry Beevers and their associates in late 195O's and early 1960's. The cycle operates in special cell organelles called *glyoxysomes,* which are small peroxisomes like bodies quite abundant in germinating fatty seeds. Hence the cycle is called glyoxylate cycle.

To begin with the process, acetyl CoA derived from the mitochondrial fatty acid oxidation (beta oxidation), diffuses to the glyoxysomes. There it reacts with oxaloacetate in the presence of enzyme citrate synthetase to form citrate, just as in TCA cycle. Citrate is then converted to isocitrate. It is then cleaved by an enzyme unique to this cycle, isocitrate lyase. A four carbon compound succinate and a two carbon compound glyoxylate is produced. The succinate moves out of the glyoxysomes into the mitochondria, where it is sequentially converted to oxaloacetate through fumarate and malate. The oxaloacetate is converted to phosphoenol pyruvate and ultimately to glucose by the reversal of glycolysis.

The glyoxylate produced in the glyoxysomes (step 3) reacts with another acetyl CoA to produce malate and free coenzyme A (CoA). This step is catalysed by another enzyme restricted to the pathway, the malate synthetase. The malate is then oxidized to oxaloacetate. This oxaloacetate in the glyoxysomes reacts with further acetyl CoA the keep the glyoxylate cycle going on.

The net result of the glyoxylate cycle is the conversion of 2 molecules of acetyl CoA (activated acetate) to succinate, which may become a substrate for the production of sugars. Had the cycle not operated, isocitrate produced in the second step would have been converted to malate via succinate and fumarate, during which two carbons are lost as CO_2. In the glyoxylate cycle, these CO_2 producing reactions are bye passed and the two carbons of acetyl CoA are conserved.

It is important to note that animals are unable to convert fatty acids into sugars.

Synthesis of Fat

Fats are synthesized from fatty acids and glycerol by the reversal of its hydrolytic process;

Synthesis of Fatty Acids: Fatty acid synthesis takes place in the cytosol at the endoplasmic reticulum. The synthesis starts with the carboxylation (by bicarbonate) of acetyl CoA to malonate;

$$\underset{\text{Acetyl CoA}}{CH_3CO{-}S{-}CoA} + CO_2 + ATP \xrightarrow{\text{Bicarbonate}} \underset{\text{Malonyl CoA}}{COOH.CH_2{-}CO{-}S{-}CoA} + ADP + Pi$$

The reaction is catalysed by the enzyme acetyl CoA carboxylase, which contains a biotin prosthetic group.

In the next step, malonyl CoA is acylated several times to produce saturated fatty acids. These reactions are catalysed by a multi enzyme complex, *fatty acid synthetase* complex. The complex is believed to be containing 6 enzymes each catalysing one step. All intermediate products remain attached to the complex. The sequential reactions are as follows:

(i) Acetyl CoA + ACP (acyl carrier protein) $\rightleftarrows$ Acetyl-ACP + CoA

(ii) Malonyl CoA + ACP $\rightleftarrows$ Malonyl-ACP + CoA

(iii) Acetyl-ACP + malonyl-ACP $\rightarrow$ Acetoacetyl-ACP + ACP + CO_2

(iv) Acetoacetyl-ACP + NADPH + H^+

$\rightleftarrows$ 3-hydroxy butyryl-ACP + NADP+

(v) 3-hydroxybutyryl-ACP $\rightleftarrows$ Crotonyl-ACP + H_2O

(vi) Crotonyl-ACP + NADPH + H^+ $\rightarrow$ Butyryl-ACP + NADP

This process of chain elongation is repeated until palmitoyl-ACP is formed. Palmitate is finally liberated from the enzyme complex.

The overall reaction for the synthesis of palmitate is as follows;

8 Acetyl CoA + 7ATP + 14NADPH

$\rightarrow$ Palmitate + $14NADP^+$ + 8CoA + $6H_2O$ + 7ADP + Pi

The high NADPH and ATP requirement indicates that fat synthesis is an expensive process. Most of the NADPH comes from presumably the pentose phosphate pathway. The palmitate thus produced is generally linked with the ACP. The palmitoyl-ACP is then converted by chain lengthening process with the help of a specific condensing enzyme to stearoyl-ACP. Most of the plant fatty acids are palmitate or stearate. For further elongation, these fatty acids migrate to endoplasmic reticulum where longer chains are formed by the activity of elongation systems.

The sequential desaturation of stearate produces unsaturated fatty acids such as oleic, linoleic and linolenic acids. Desaturases, catalysing these reaction, have been found to be present in plastids.

Fatty Acid Synthetase: The enzyme fatty acid synthetase isolated from eukaryotes is in fact an aggregate of several enzymes which catalyse individual steps in the biosynthesis of saturated fatty acids starting from malonyl CoA. The enzyme therefore is referred to as fatty acid synthetase complex. It has been isolated, purified and characterised from a variety of organisms including plants. Depending upon the tightness of the component units, the fatty acid synthetase complex are of two types: type I and type II.

Type I complex contains all closely linked units, which aggregate and purify together. There are two types of type I complex: IA and IB. IA fatty acid synthetase complex has a m.w. of about 500 kD and its product is free palmitic acid. IB complex has a m.w. in the range of 1000 kD to 2300 kD and its product is a acyl CoA derivative.

In type II complex, the component units are loosely associated and they can be readily separated. Most plant synthetases are of this type.

Several factors are known to regulate the activity of fatty acid synthetase complex. Metabolites, fructose bisphosphate and palmitic acid have opposite effects on enzyme activity. In pigeon liver, fatty acid synthetase is stimulated by 10-30 mM fructose 1,6-bisphosphate and possibly by other phosphorylated sugars as well. The sugars lower the Km NADPH of the enzyme by almost four folds. The enzyme from rat liver is activated by ATP and Mg^{2+}. In many other mammalian systems, the enzyme is reduced by insulin or carbohydrate diet after starvation. The fatty acid synthetase complex from the yeast is inhibited by palmitoyl CoA. This inhibition is reversible.

The plant fatty acid synthetase complex was purified by three groups of scientists working independently in 1982. As mentioned earlier, it is of II type. The complex catalyses the production of palmitoyl-ACP and stearoyl-ACP.

Synthesis of Glycerol: Glycerol is synthesized from dihydroxyacetone phosphate as glycerol phosphate through glyceraldehyde phosphate;

Dihydroxyacetone phosphate $\rightarrow$ glyceraldehyde phosphate $\rightarrow$ glycerol phosphate

The reaction is catalysed by NADH dependent glycerol 3-phosphate dehydrogenase.

Synthesis of Triacylglycerols (Fats): The biosynthesis of triacylglycerols occurs primarily on the surface of endoplasmic reticulum. Three fatty acid molecules join sequentially to a glycerol phosphate molecule to form a triacylglycerol.

The fatty acids are first activated by their conversion to respective acyl derivatives by the enzyme Acyl CoA synthetase;

$$R\text{—}COOH + CoASH + ATP \xrightarrow{\text{Acyl CoA synthetase}} R\text{—}\overset{\overset{\displaystyle O}{\|}}{C}\text{—}S\text{—}CoA + AMP + PPi$$

The esterification of glycerol phosphate with this fatty acyl CoA (RCO—SCoA) takes place in three successive steps;

(i)

$$\begin{array}{l} CH_2OH \\ | \\ CHOH \\ | \\ CH_2O(P) \end{array} + R_1\text{—}CO\text{—}SCoA \xrightarrow[\text{CoASH}]{\text{Glycerol phosphate acyl transferase}} \begin{array}{l} CH_2OH \\ | \\ CHO\text{—}\underset{\underset{\displaystyle O}{\|}}{C}\text{—}R_1 \\ | \\ CH_2O(P) \end{array}$$

Monoacyl glycerol phosphate or Lysophosphatidate

(ii)

$$\begin{array}{l} CH_2OH \\ | \\ CHO\text{—}\underset{\underset{\displaystyle O}{\|}}{C}\text{—}R_1 \\ | \\ CH_2O(P) \end{array} + R_2\text{—}CO\text{—}SCoA \xrightarrow[\text{CoASH}]{\text{Lysophosphatidate acyl transferase}} \begin{array}{l} CH_2O\text{—}\underset{\underset{\displaystyle O}{\|}}{C}\text{—}R_2 \\ | \\ CHO\text{—}\underset{\underset{\displaystyle O}{\|}}{C}\text{—}R_1 \\ | \\ CH_2O(P) \end{array}$$

Monoacyl glycerol phosphate or Lysophosphatidate → Diacyl glycerol phosphate or Phosphatidate

(iii)

$$\begin{array}{l} CH_2O\text{—}\underset{\underset{\displaystyle O}{\|}}{C}\text{—}R_2 \\ | \\ CHO\text{—}\underset{\underset{\displaystyle O}{\|}}{C}\text{—}R_1 \\ | \\ CH_2O(P) \end{array} + R_3\text{—}CO\text{—}SCoA \xrightarrow[+H_2O \quad \text{CoASH}]{\text{Diacyl glycerol acyl transferase}} \begin{array}{l} CH_2O\text{—}\underset{\underset{\displaystyle O}{\|}}{C}\text{—}R_2 \\ | \\ CHO\text{—}\underset{\underset{\displaystyle O}{\|}}{C}\text{—}R_1 \\ | \\ CH_2O\text{—}\underset{\underset{\displaystyle O}{\|}}{C}\text{—}R_3 \end{array} + Pi$$

Diacyl glycerol phosphate or Phosphatidate → Triacyl glycerol

The enzymes carrying out acylations of glycerol phosphate are part of a membrane bound triacyl glycerol synthetase complex.

Nucleic Acid Metabolism

The nucleic acids, like proteins, are macromolecules. They are made up of usually four types of nucleotides. The nucleotides contain (i) a purine or pyrimidine base (ii) a ribose or deoxyribose sugar and (iii) a phosphate group.

Pyrimidine and Purine Bases: Pyrimidine is an aromatic six membered ring that contains two nitrogen atoms at position 1 and 3;.

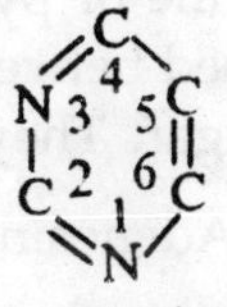

pyrimidine

The most abundant pyrimidines are uracil (U) found in ribonucleic acid (RNA); cytosine (C) in both RNA and deoxyribonucleic acid (DNA) and thymine (T) found in DNA;

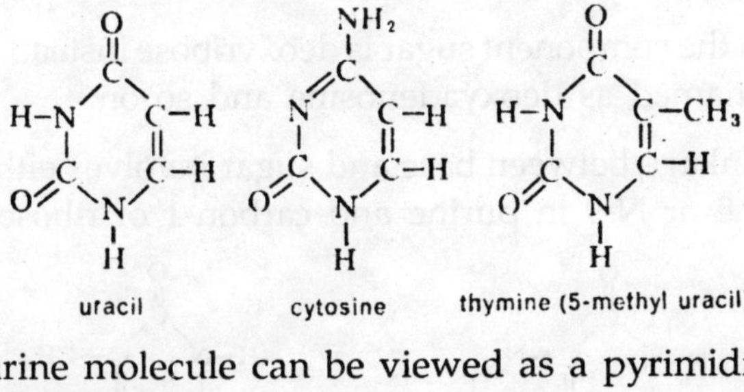

uracil cytosine thymine (5-methyl uracil)

A purine molecule can be viewed as a pyrimidine fused with an imidazole;

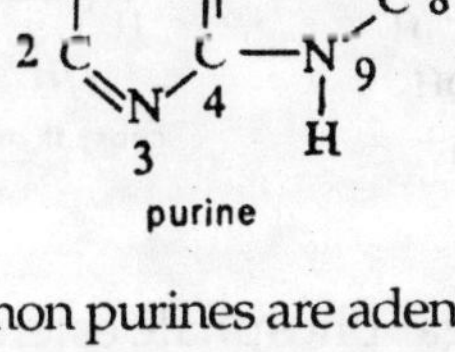

purine

The two most common purines are adenine (A) and guanine (G):

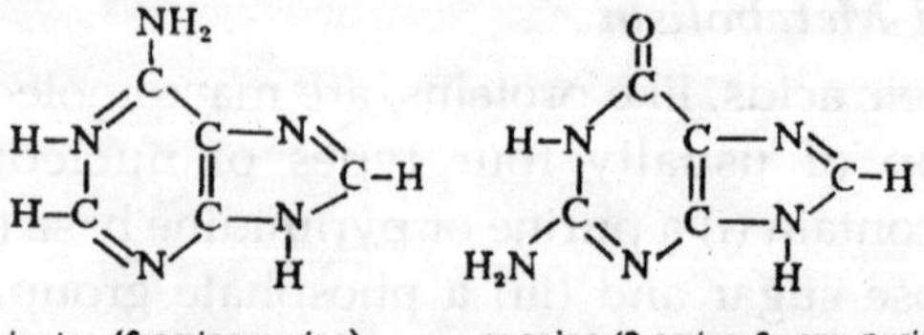

adenine (6-amino purine) guanine (2-amino 6-oxy-purine)

Nucleosides

Nucleosides are second step in the formation of the macromolecule from purine or pyrimidine base and a sugar. They are glycosides formed between one of the bases and the ribose or deoxyribose sugar. They are named as follows:

Adenine + ribose = Adenosine

Guanine + ribose = Guanosine

Uracil + ribose = Uridine

Cytosine + ribose = Cytidine

Thymine + ribose = Thymidine

When the component sugar is deoxyribose instead of ribose, they are named as deoxyadenosine and so on.

The linkage between base and sugar involves either N-l in pyrimidine or N-9 in purine and carbon-1 of ribose sugar,

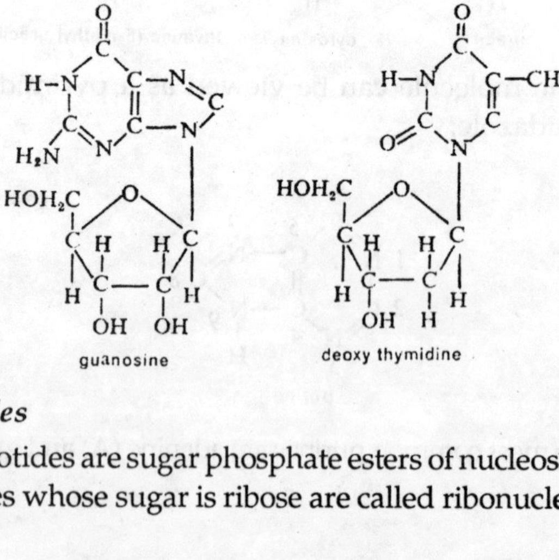

guanosine deoxy thymidine

Nucleotides

Nucleotides are sugar phosphate esters of nucleosides. Those nucleotides whose sugar is ribose are called ribonucleotides and those derived from 2-deoxyribose are known as deoxyribonucleotides. Each of the free hydroxyl groups of ribose can be phosphorylated, and 2'-, 3'- and 5'-phosphate esters of ribonucleotides are known.

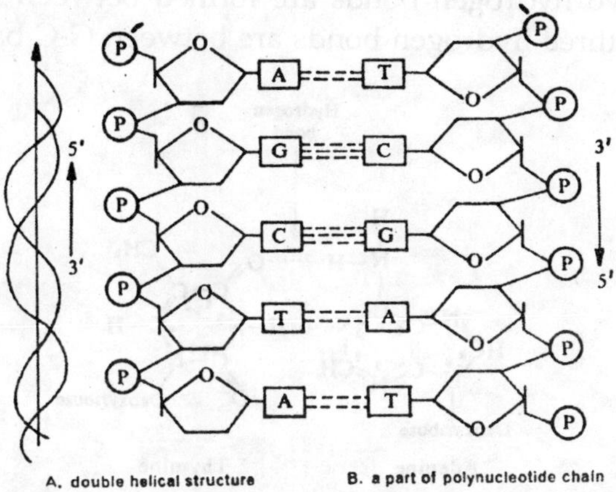

A. double helical structure B. a part of polynucleotide chain

Diagrammatic representation of the structure of DNA (A) Double helical structure (B) A part of polynucleotide chain.

Nucleic acids are constructed from the nucleotides. Nucleotides join by phosphodiester bonds between the 3' and 5' positions of ribosyl moieties. The primary structure of either DNA or RNA is the linear sequence of nucleotide residues. Thus, it is either a polydeoxyribonucleotides or a polyribo- nucleotides.

Deoxyribonucleic Acid

In the eukaryotic cell, the DNA is present mainly within the nucleus and is termed chromatin. DNA is also present in the chloroplasts and mitochondria.

The DNA is made up of two polynucleotide chains (double stranded). It consists of about 8000 to 15000 base pairs in length and thus has a molecular weight in the range of 5×10^6 to 10^7 or sometimes upto 10^{10}. The two chains of DNA are twisted and interlocked in the form of a right handed helix. One chain runs with its phosphodiester bonds advancing in the 3' → 5' direction, while the other is anti-parallel with its phosphodiester bonds running in 5' → 3' direction. The phosphate groups of nucleotides form the outer surface of the helix, while purine and pyrimidine bases are on the inner side, the two chains are held together by hydrogen bonds formed between A and T and between G and C. Two hydrogen bonds are formed between A-T base pairs and three hydrogen bonds are between G-C base pairs.

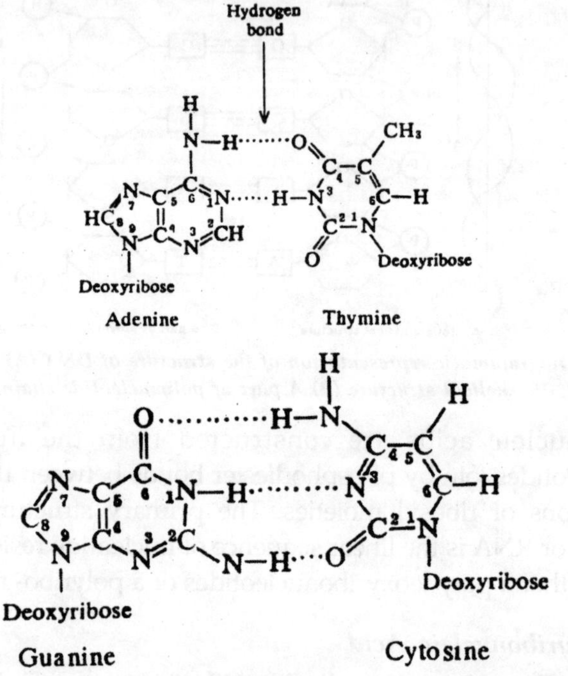

Hydrogen bond

Deoxyribose Deoxyribose

Adenine Thymine

Deoxyribose Deoxyribose

Guanine Cytosine

Base pairing of adenine and thymine and guanine and cytosine.

Chloroplastic DNA: Chloroplasts of several plant species have been found to contain large, circular and supercoiled DNA. The length and the molecular weight of the molecule varies with the species.

One chloroplast may contain between 20 and 60 copies of DNA molecule. Some ribonucleotides (instead of usual deoxyribonucleotides) have also been discovered in chloroplastic DNA. Possibly they play a role in DNA replication.

Plant Mitochondrial DNA: Plant mitochondrial DNA is a circular molecule with a molecular weight in the range of 220 to 166 × 10^6. It is about seven times bigger than the animal mitochondrial DNA. It is responsible for the synthesis of most of the mitochondrial proteins including enzymes.

In contrast to nuclear DNA, the DNA from mitochondria and chloroplasts is devoid of histone proteins.

Ribonucleic Acid

RNA consists of a long unbranched polynucleotide chain in which the adjacent sugar residues are joined by 3' → 5' phosphodiester bonds, just like in DNA. As in other organisms, plants also contain 3 types of RNA depending upon their role : (i) messenger RNA (ii) transfer RNA and (iii) ribosomal RNA.

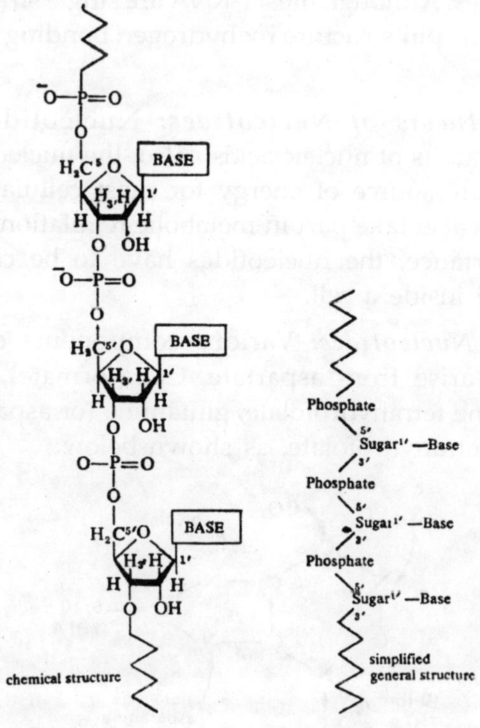

Structure of part of a RNA molecule.

Messenger RNA (m-RNA) molecules carry genetic information from DNA, for protein synthesis. Their nucleotide sequence is similar to that of one strand of DNA and this sequence determines the amino acid sequence of proteins. Although the m-RNA in most organisms is single stranded, it may assume secondary folding in many species.

Transfer or soluble RNA (t-RNA or s-RNA) makes about 10 to 20 percent of the cellular RNA and is comparatively smaller in size. It is made up of 75 to 90 nucleotides. The molecule transfers amino acids from cytoplasmic pool to the ribosomes for protein synthesis. The t-RNA has specific secondary and tertiary structure.

Ribosomal RNA (r-RNA) may comprise upto 80% of the total cellular RNA. Possibly it assists in protein synthesis on the ribosomes. Although most r-RNA are single stranded, some assume a hair pin structure by hydrogen bonding in the same chain.

Biosynthesis of Nucleotides: Nucleotides are the monomeric units of nucleic acids. Also, the nucleotide ATP is an immediate source of energy for most cellular processes. Nucleotides also take part in metabolic regulation. Because of their importance, the nucleotides have to be continuously synthesized inside a cell.

Purine Nucleotides: Various components of a purine nucleotide arise from aspartate, CO_2 (formate), glycine, 5, 10-methylene tetrahydrofolate, glutamine (or asparagine) and 10-formyl tetrahydrofolate, as shown below:

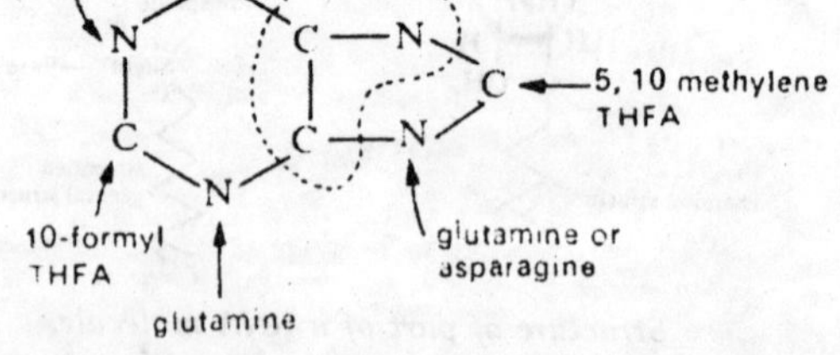

CO_2
aspartate
glycine
5, 10 methylene THFA
10-formyl THFA
glutamine or asparagine
glutamine

In most organisms, the pathway of purine nucleotide biosynthesis consists of 11 steps.

In the initial reaction, ribose 5-phosphate is phosphorylated by ATP to produce 5-phosphoribosyl 1-pyrophosphate (PRPP). The PRPP is then aminated by either glutamine or asparagine, which seems to be the most efficient amino donor in this reaction in plant cells. A pyrophosphate group is lost and 5-phosphoribosyl amine (PRA) is produced. In the next step, the amino group of the PRA molecule reacts with the carboxyl

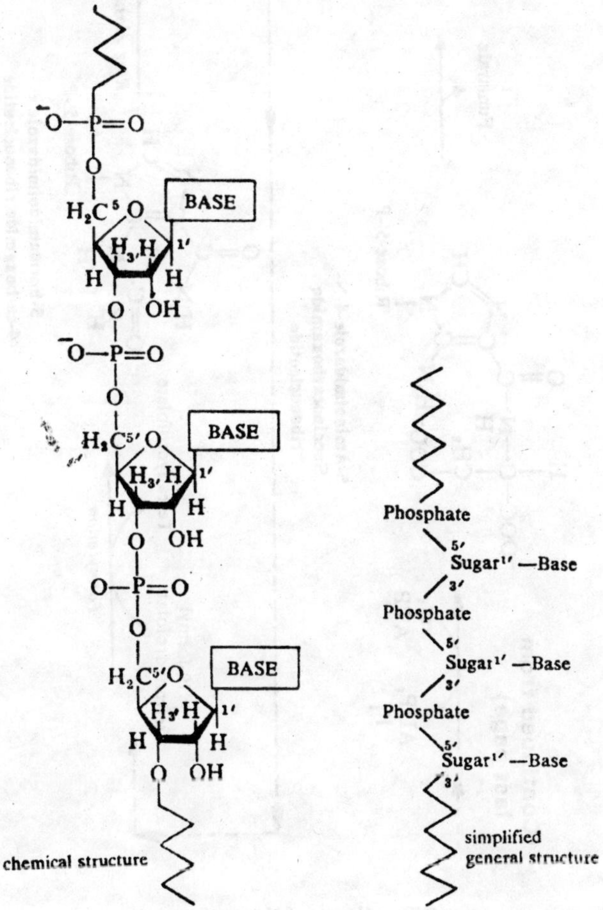

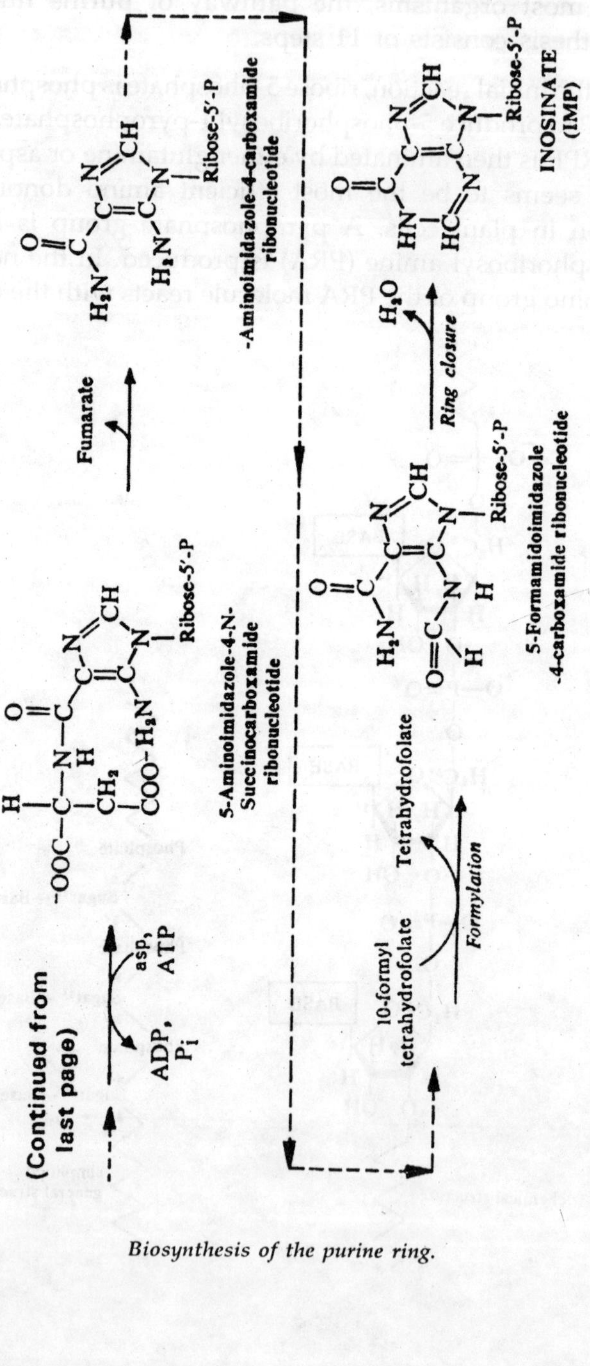

(Continued from last page)

ADP, P_i

asp, ATP

5-Aminoimidazole-4-N-Succinocarboxamide ribonucleotide

Fumarate

-Aminoimidazole-4-carboxamide ribonucleotide

10-formyl tetrahydrofolate

Tetrahydrofolate

Formylation

5-Formamidoimidazole -4-carboxamide ribonucleotide

H_2O

Ring closure

INOSINATE (IMP)

Biosynthesis of the purine ring.

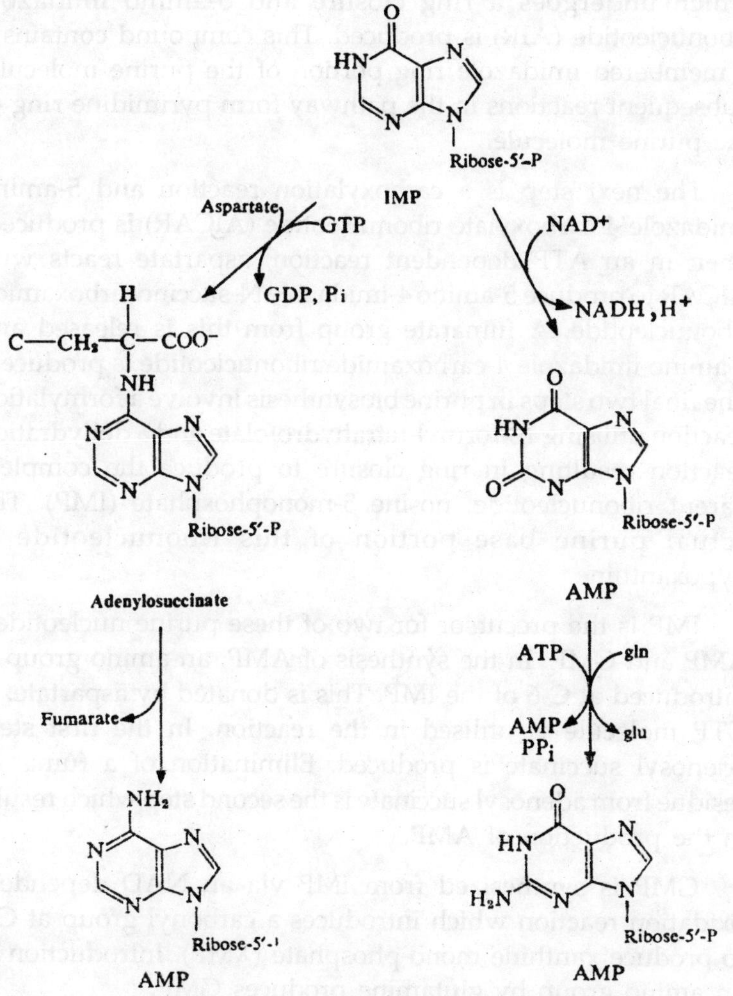

Conversion of IMP to adenine and guanine nucleotides.

group of glycine to form an amide link between glycine and the PRA and to produce glycinamide ribonucleotide (GAR). GAR is then formylated by methylene tetrahydrofolate to yield formyl glycinamide ribonucleotide (FGAR). In the next step, FGAR undergoes amination, in which the amide nitrogen of glutamine is used up to produce an amidine group in FGAR. As a result fortmyl glycinamide ribonucleotide is produced which undergoes a ring closure and 5-amino imidazole ribonucleotide (AIR) is produced. This compound contains a 5 membered imidazole ring portion of the purine molecule. Subsequent reactions in the pathway form pyrimidine ring of the purine molecule.

The next step is a carboxylation reaction and 5-amino imidazole 4-carboxylate ribonucleotide (AICAR) is produced. Then in an ATP dependent reaction, aspartate reacts with AICAR to produce 5-amino 4-imidazole N-succinocarboxamide ribonucleotide. A fumarate group from this is released and 5-amino imidazole 4-carboxamide ribonucleotide is produced. The final two steps in purine biosynthesis involve a formylation reaction utilising 10-formyl-tetrahydrofolate and a dehydration reaction resulting in ring closure to produce the complete parent ribonucleotide, nosine 5-monophosphate (IMP). The actual purine base portion of this ribonucleotide is hypoxanthine.

IMP is the precursor for two of these purine nucleotides, AMP and GMP. In the synthesis of AMP, an amino group is introduced at C-6 of the IMP. This is donated by aspartate. A GTP molecule is utilised in the reaction. In the first step, adenosyl succinate is produced. Elimination of a fumarate residue from adenosyl succinate is the second step which results in the production of AMP.

GMP is synthesized from IMP via an NAD dependent oxidation reaction which introduces a carbonyl group at C-2 to produce xanthine mono-phosphate (XMP). Introduction of an amino group by glutamine produces GMP.

Pyrimidine Nucleotides

In contrast to purine biosynthesis, where the purine ring is formed while linked to ribose 5-phosphate (*i.e.*, via nucleotide intermediate), the pyrimidine ring is synthesized first and then linked to ribose 5-phosphate to form a nucleotide. Various components of pyrimidine are derived from aspartate, glutamine and CO_2.

Pyrimidine biosynthesis begins with the formation of carbamoyl phosphate, utilizing glutamine, bicarbonate and ATP. The reaction is catalysed by the enzyme carbamoyl phosphate synthetase. In the next step, carbamoyl phosphate condenses with aspartate in the presence of enzyme aspartate transcarbamoylase to produce N-carbamoyl aspartate. Then carbamoyl aspartate is dehydrated to form dihydro-orotate, which in turn is oxidised in an NAD dependent reaction to produce orotate. Orotate is the first true pyrimidine derivative in the pathway.

Orotate reacts with phosphoribosyl pyrophosphate (PRPP) to form a pyrimidine nucleotide orotidylate. The pyrophosphate is liberated in a reaction. Orotidylate is decarboxylated to produce uridylate (uridine monophosphate, UMP). UMP can be converted to UDP and UTP by Phosphorylation and ultimately to CTP by amination with glutamine.

DNA Replication

As in other organisms, the DNA molecule is replicated semi-conservatively in plants also. Each of the two strands of a double stranded DNA, acts as a Opiate for the synthesis of a new complimentary strand. This results in the formation of two molecules of double stranded DNA with each molecule consisting of one old (parental) and one new (daughter) DNA strand . Some essential features of DNA replication are as follows:

(1) Both strands may be replicated simultaneously in opposite direction ($5' \rightarrow 3'$ for each strand).

(2) The replication is semi discontinuous. Small fragments are first formed, which join later on to form large DNA strands.

(3) The RNA primers are formed before the formation of DNA fragments.

Mechanism

The actual process of DNA replication can be divided into five stages

(i) Unwinding of the parental double helix,

(ii) Synthesis of a oligonucleotide primer of RNA,

(iii) Formation of DNA chain in 5' $\rightarrow$ 3' direction,

(iv) Excision of the RNA primer and

(v) Joining of the newly synthesized DNA chains by filling in the excision gap and sealing by phosphodiester formation.

The replication starts at a specific initiation point, which is recognised by an initiation protein. Several proteins are involved in unwinding of the double helical DNA also. Some proteins bind the separated strands of DNA and keep them separated for replication. Only a limited portion of DNA is separated and a kind of bubble or loop is formed in the DNA molecule. One of the two strands acts as a template. The synthesis of DNA on this template begins at a specific site with the formation of a RNA primer of 8-10 nucleotides by the enzyme RNA polymerase. On this primer a small segment of DNA (100 to 200 nucleotides) is formed by using deoxyribonucleotides.

The enzyme involved is DNA polymerase-α in eukaryotes. The RNA primer is then removed by either exonuclease activity of one of the DNA polymerases or by ribonuclease H. The small fragments of DNA thus left are called *Okazaki fragments.* Later, the gap between ukazaki fragments is filled up with the deoxyribonucleotides by the action of DNA polymerase-α. The incorporation of final nucleotide in the gap require in addition to DNA polymerase-α, other essential cofactors also. When the final nucleotide has been inserted, DNA ligase I joins the two ends of adjacent polynucleotide chains together by formation of a phosphodiester linkage.

DNA replication in plants and other eukaryotic cells appears to occur in stages from smaller fragments to larger fragments of DNA. First 40-290 nucleotide long Okazaki fragments are formed. Then these fragments are joined to form pieces containing 500 to 800 nucleotides. Then pieces of 1000 to 4000 nucleotides are formed. The formation of larger and larger pieces eventually leads to replication of entire chromosome without the formation of discrete intermediates of varying sizes.

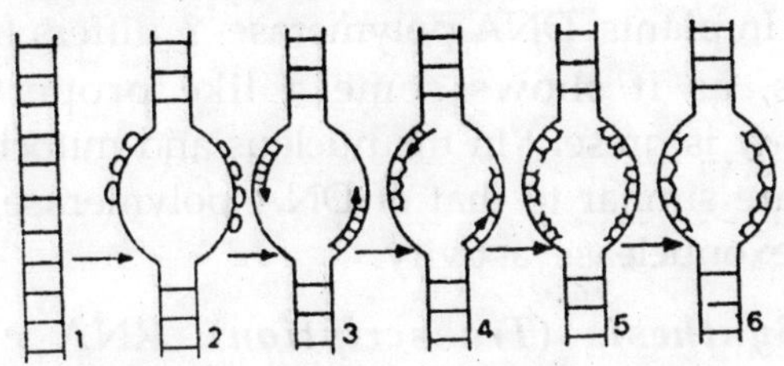

Diagrammatic representation of stages in the replication of DNA.

1. Recognition of initiation point by initiator protein.
2. Dissociation of H bond and opening up of a bubble in the duplex by unwinding proteins.
3. Formation of RNA primers by RNA polymerase on two open strands.
4. Formation of DNA strands on RNA pruners by poly-III-copoly III complex.
5. Removal of RNA primers and formation of Okazaki fragments.
6. Filling up of the gaps between Okazaki fragments and formation of DNA strand.

DNA Polymerases

Many proteins are involved in eukaryotic DNA replication. Perhaps the most extensively studied are DNA polymerases. Four types of DNA polymerases—α, β, γ and δ have been identified.

DNA polymerase-α is the enzyme solely responsible for DNA replication in the nucleus. It is present inside the nucleus and the activity of the enzyme is often correlated with the rate of cellular DNA replication.

DNA polymerase-β is also present inside the nucleus. It is responsible for DNA repair in the nucleus. It adds single nucleotides to an activated DNA primer.

DNA polymerase-γ is found in nucleus and mitochondria. Its function in nucleus is not known. However, in the mitochondria it is responsible for mitochondrial DNA replication. In plants, DNA polymerase-γ differs from that in vertebrates, as it shows some a like properties. DNA polymerase-δ is present in the nucleus and mitochondria. Its properties are similar to that of DNA polymerase, but it has associated exonuclease activity.

RNA Synthesis (Transcription): RNA chains are synthesized on DNA template in 5' $\rightarrow$ 3' direction, just like DNA chains. RNA polymerase catalyses this reaction. The enzyme binds at a specific place on DNA called the promoter site. The promoter site has a specific sequence and in the absence of such a sequence the transcription of the sequence is greatly diminished. In both prokaryotes and eukaryotes, TATA also called as 'Goldberg-Hogness' box is located 25-30 nucleotides upstream from the start of transcription site. A short unwinding or 'melting' of the duplex occurs to provide access to the DNA template. RNA polymerase then copies the DNA chain in 3' $\rightarrow$ 5' direction thus synthesizing the RNA chain in 5' $\rightarrow$ 3' direction. In other words, during transcription process first 5' terminus of the cap site of RNA if formed and 3' terminus is formed in the end. The DNA strand which is transcribed to produce RNA chain is called 'sense strand'.

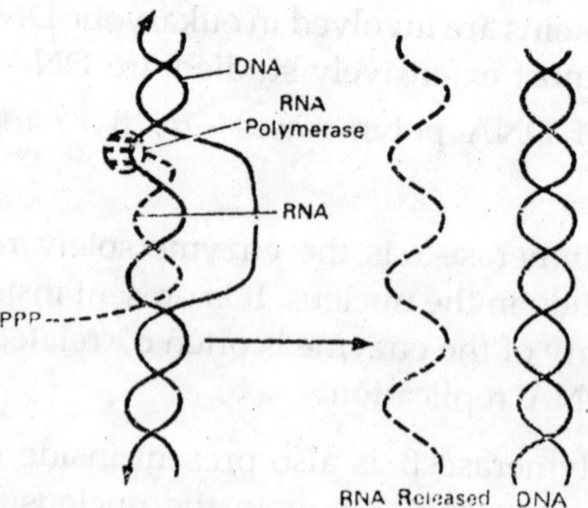

Diagrammatic representation of transcription of DNA by RNA polymerase.

The enzyme RNA polymerase adds ribonucleotide monophosphate units in sequence. The order of the addition of ribonucleotides is determined by the base sequence of the template strand. Simultaneously with the formation of RNA chain, some bases of the ribonucleotide are chemically modified. The chemical modification may take place later on as well. Synthesis of RNA stops at a particular site on the DNA molecule. It is suggested that some stop signals are present in DNA molecule. The synthesized RNA is released with the help of a release factor and processed for its specific role.

RNA Polymerases

Plant cells contain three types of RNA polymerases in the nucleus; RNA polymerase I, II and III. Chloroplastic and mitochondrial RNA polymerases are also known, but there is very little information about their properties.

RNA Polymerase I is present in the nucleolus. It is eluted with 0.1 M ammonium sulfate from a DEAE-sephadex column and its activity is unaffected by α-amanitin, a poisonous mushroom toxin. This enzyme is responsible for the formation of r-RNA (28 S and 18 S).

RNA Polymerase II is located in the nucleoplasm. It is eluted from the chromatography column with 0.2 M ammonium sulfate and is inhibited by very low concentrations of α-amanitin. The enzyme is responsible for the synthesis of heterogeneous m-RNA (precursor of m-RNA).

RNA Polymerase III is located in the nucleoplasm. It is eluted with 0.3 M ammonium sulfate and is inhibited by relatively high concentrations of a-amanitin. The enzyme is responsible for the synthesis of low molecular weight RNA such as t-RNA and 5 S RNA.

Each of the eukaryotic RNA polymerase is a large and complex protein, with several sub units of variable sizes. Molecular weight of sub units of some plant RNA polymerases are given in table.

Structure and Molecular Weight of Sub Units of Some Plant RNA Polymerases

Source	*RNA Polymerase*	*m.w. (× 10^3) of the sub units*
Soyabean	Polymerase I (7 sub units)	183, 136, 50, 46, 40, 33, 28
	Polymerase II (8 sub units)	170, 142, 42, 28, 20, 16, 15.5, 14
Maize	Polymerase II (6 sub units)	220, 160, 35, 25, 20, 17
Wheat embryo	Polymerase I (7 sub units)	200, 125, 38, 24, 20, 17.8, 17
	Polymerase II (11 sub units)	220, 140, 82, 52, 21, 20, 17.8, 17, 16.3, 16, 14
	Polymerase III (10 sub units)	150, 130, 100, 55, 38, 28, 25, 20, 17.8, 17

Heterogeneous RNA and Processing of m-RNA: In plants and other eukaryotes, m-RNA formed as a result of DNA transcription is quite different than the m-RNA, which actually participates in protein synthesis. It is much bigger (1500 to 3000 nucleotides) in size and is called heterogeneous nuclear RNA (Hn RNA) or high molecular weight RNA (HMW RNA). It undergoes several types of chemical modifications. Generally a sequence of nucleotides is removed from both ends of the RNA. After that a 7-methyl guanylate residue is added to the 5' end and a polyadenylate tail of 150 to 200 nucleotides is added to the 3' end. An enzyme capable of polyadenylate synthesis has been found in many plant tissues. A few m-RNA such as histone m-RNA do not contain a polyadenylate tail. The function of polyadenylate tail in m-RNA is not known. The presence of 7-methyl guanylate residue at 5' end of the m-RNA is designated as cap 0. Other nucleotides in this capped m-RNA can also be methylated. Addition of one more methyl group (besides 7-methyl guanylate at 5' end) gives a product known as cap I and addition of one more methyl gives rise to cap II. A cap stabilizes m-RNA by protecting it from exonuclease and phosphatases. Possibly it helps in the recognition of ribosome also.

In the final processing of m-RNA, some splitting and rejoining of various segments takes place to form the final m-RNA. Splitting is catalysed by endonucleases and joining by ligases.

Catabolism of Nucleic Acids

Like other macromolecules, nucleic acids are also degraded enzymatically to yield micromolecules, such as nucleotides and other simpler components. This type of metabolism is quite apparent in senescing plant tissues. The products are then metabolised and transported to younger tissues, where they are utilised in rebuilding of nucleic acids and other nitrogenous molecules.

The most important type of nucleic acid catabolism is the hydrolysis of phosphodiester bonds in a polynucleotide chain, by the enzymes called nucleases. Nucleases are classified into different ways. The most elementary distinction is based on whether they act on DNA or RNA. Deoxyribonucleases (DNAases) are specific for DNA while ribonucleases (RNAases) are for RNA. However, a few nucleases are able to hydrolyses both DNA and RNA.

According to another criterion, *i.e.*, point of attack in a nucleic acid chain the nucleases are classified into exonucleases and endonucleases. Exonucleases attack the DNA or RNA chain at either 3' or 5' terminus and sequentially remove nucleotidyl residue one at a time. Endonucleases attack the DNA/RNA within the chain. Endonucleases hydrolyse and remove nucleotidyl residue from within the chain.

The nucleases hydrolyse nucleic acids to yield either 5' nucleotide or 3' nucleotides. Those nucleases that yield 5' nucleotides monophosphates as products catalyse a-type cleavage and those that yield 3' nucleotide monophosphates catalyse b-type cleavage. In other words, a-type nucleases start cutting off nucleotides from 3' terminus of DNA or RNA while b-type enzyme starts from 5' terminus of the nucleic acid.

Catabolism of Nucleotides

Nucleotides derived from the enzymic hydrolysis of nucleic acids are further degraded. In animals, a major portion of their degradative products is reutilised in other metabolic pathways and the remainder is excreted out. Generally nucleotidases degrade nucleotides to nucleosides by hydrolysis and then nucleosidases catalyse the phosphorylytic cleavage of nucleosides to free bases and ribose 1-phosphate.

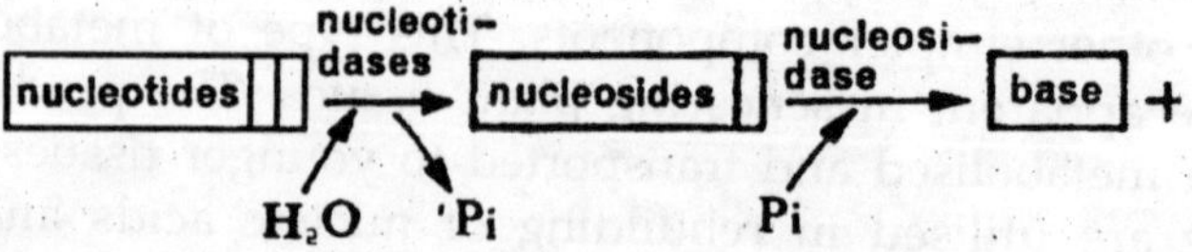

Diagrammatic representation of the hydrolysis of nucleotides.

$$\text{Nucleotides} \xrightarrow[H_2O \quad Pi]{\text{Nucleotidases}} \text{Nucleosides} \xrightarrow[Pi]{\text{Nucleosidases}} \text{Base + Ribose 1-P}$$

The bases can be metabolised further. In humans, primates and some reptiles, purine bases are converted to uric acid. The uric acid can be further metabolised to produce products such as allantoin, urea and ammonia. These Products can be excreted out.

The pyrimidine bases are metabolised through uracil. Pyrimidine cytosine is deaminated to produce uracil. The uracil is then reduced to N-carbamoyl propionate. In the final step, N-carbamoyl propionate is hydrolysed to yield β-alanine, NH_4^+ and CO_2.

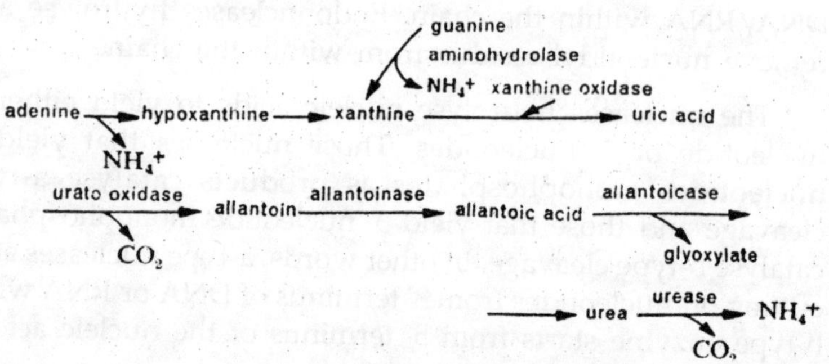

Diagrammatic representation of purine catabolism.

cytosine → uracil → dihydro-uracil → N-carbamoyl propionate

NH_4^+

→ β-alanine + NH_4^+ + CO_2

Diagrammatic representation of pyrimidine catabolism.

The general route of nucleotide degradation in plant is also similar.

Protein Structure

Proteins are the most important chemicals of living cells. They perform a variety of functions such as; building the cell wall and membrane structure, assisting in the transport of smaller molecules across the membranes, acting as storage form of nitrogen in the seeds and catalysing the biochemical reactions. The catalytic function of proteins is the most important and dynamic aspect of their existence. It is believed that there are no chemical reactions that take place in living cells, that are not regulated by one or more enzymic proteins.

Structure and Classification of Proteins: Proteins are large molecules with molecular weight ranging between 6 and 200 kDa. All proteins are made up of 20 or less types of amino acids. The amino acids have at least one carboxylic (—COOH) group and an amino (—NH_2) or imino (—NH) group. The amino/imino group is attached to the alpha carbon atom, *i.e.*, carbon next to COOH group carbon. As far as is known, all the plant protein amino acids have 1-configuration. The structural formula of these amino acids are given in figure. Besides, there are about 100 non-protein amino acids.

In the formation of proteins, amino acids are linked through a covalent bond between the alpha amino group of one amino acid and carboxylic group of the other. A water molecule is eliminated. Such a bond is called *peptide bond*. The resulting dimer, "made up of two amino acids, is called a dipeptide. Three amino acids bonded together through peptide bonds are called a tripeptide. Similarly four amino acids link linearly to form tetrapeptide, and so on. Long chain polymers of amino acids are usually termed polypeptides. Polypeptide is the back bone of proteins; it forms the primary structure of protein.

The polypeptides fold and refold to form stable two dimensional and three dimensional configurations of proteins. However, some polypeptides function in plants just as polypeptides, without any secondary or tertiary structure. A number of glutamyl and polyglutamic acid polypeptides are known from higher plants. Glutathione is a well known peptide which functions as a cofactor for many enzymic reactions.

The proteins are classified according to their size, solubility and degree of basicity. But the classification based on their solubility is more important, as it is the basis for isolating and

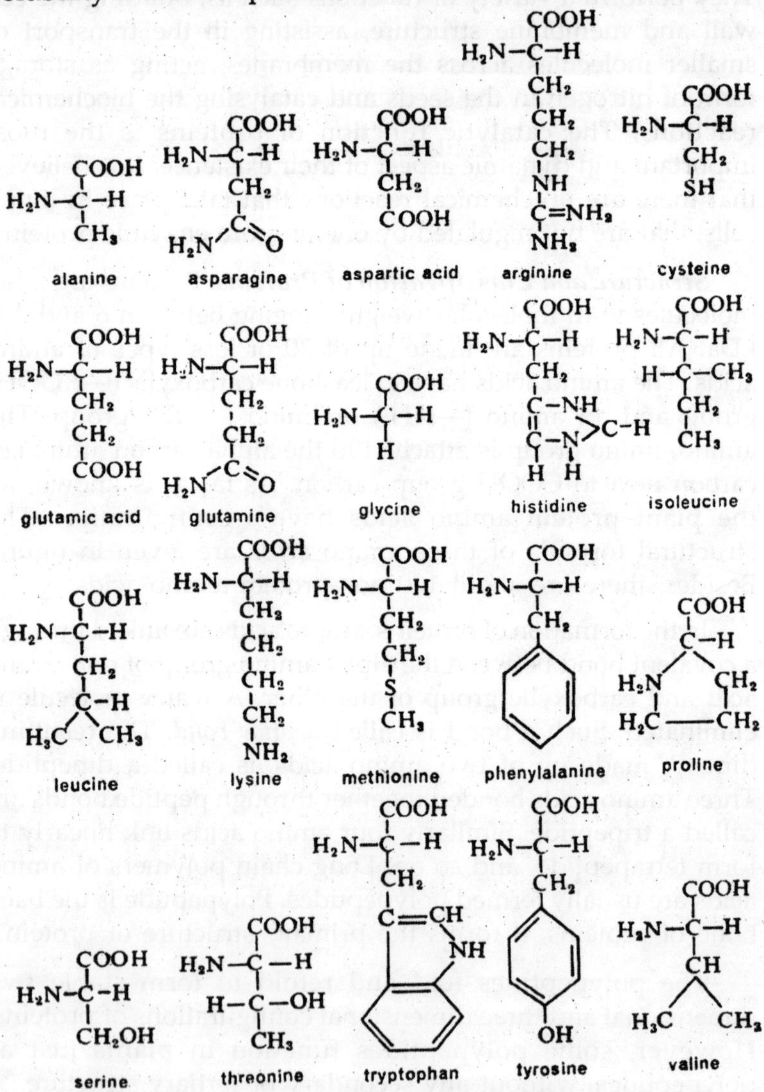

alanine asparagine aspartic acid arginine cysteine

glutamic acid glutamine glycine histidine isoleucine

leucine lysine methionine phenylalanine proline

serine threonine tryptophan tyrosine valine

Structure of protein amino acids

$$\begin{array}{c} H \\ | \\ H_2N{-}C{-}COOH \\ | \\ R_1 \end{array} + \begin{array}{c} H \\ | \\ H_2N{-}C{-}COOH \\ | \\ R_2 \end{array} \underset{H_2O}{\overset{H_2O}{\rightleftharpoons}} \begin{array}{c} H \quad O \quad \quad H \\ | \quad \| \quad \quad | \\ H_2N{-}C{-}C{-}N{-}C{-}COOH \\ | \quad \quad | \quad | \\ R_1 \quad \quad H \quad R_2 \end{array}$$

Formation of a peptide (CONH) bond between two amino acids

purifying proteins from most plant sources. Accordingly, the proteins are classified in to following 4 major groups:

1. *Albumins:* Soluble in water and in dilute salt solutions, readily precipitated by about 40% ammonium sulfate solution. These are the most important proteins of plant leaf.
2. *Globulins:* Insoluble in water, soluble in dilute salt solutions, can be precipitated by 40% ammonium sulfate solution. These are the common proteins of some seeds, such as ground nut.
3. *Glutenins:* Insoluble in water and dilute salt solutions, can be dissolved in acid and alkaline solutions. These are the common proteins of cereals.
4. *Prolamines:* Insoluble in water and salt solutions but soluble in 70-80% alcohol. These are found in seeds. Zein of maize seeds is one example of prolamines.

Protein Content of the Plants

Protein content of the plant tissue is quite variable. However, in most species the seeds contain maximum amount of protein, which is stored there as a reserve form of nitrogen. On an average, the seed protein content is in the range of about 10% (example maize) to about 30% (example ground nut) in most of cereals. The leaves contain about 2% protein on dry weight basis. The major part of the protein in the seeds is stored in oval or spherical specialized cellular organelles called *aleurone grains* or *protein bodies.* The protein bodies are about 1 to 20 μm in diameter; they are packed with protein and have a single membrane. In addition to storage protein, the protein bodies also have hydrolytic enzymes such as phosphatases, proteases and amylases and some non-protein substances such as phosphate, potassium and calcium. Protein bodies have been identified in ground nut, cotton seed, soybean, maize, wheat, rice, barley, beans and peas.

The major part of the leaf protein is located in the chloroplasts, although a smaller proportion is present in other cell organelles and cytosol as well the chlorophyll exists conjugated with proteins, the fraction I protein (Ribulose bisphosphate carboxylase-oxygenase; RUBISCO) is the major protein of the chloroplasts.

The protein content of roots is low, usually in the range of 1 to 1.5% on fresh weight basis. The content decreases as the roots mature. The stems also have a very low level of proteins.

Protein Changes During Seed Germination and Seedling Growth: The protein in the seeds (as protein bodies) may be stored either in endosperm as in monocotyledons or in cotyledons as in dicotyledons. In cereal seeds, the cells of outermost layer of endosperm, called *aleurone layer,* are particularly rich in protein bodies.

During germination, seeds imbibe water, setting off a variety of chemical reactions that lead to the emergence of the radicle. The proteins present in the protein bodies are hydrolysed by proteases or peptidases to amino acids and amides. There is some controversy as to whether the proteases are present in protein bodies in an inactive form prior to the commencement of germination or whether they are produced after onset of germination. In maize seeds, the protease activity is very low and it increases during germination. Further, the increase is checked by the inhibitors of protein synthesis; suggesting that the proteases are synthesized *de novo* during germination. Increase in protease during germination has been reported in pea, bean and rice also. Ultra-structural studies indicate that during germination, protein hydrolysis occurs uniformly throughout the protein body. Hydrolysis of protein decreases the protein content of endosperm/cotyledons during germination and increases the amino acid content.

The amino acids produced may be used to produce other enzymic proteins in the storage tissue itself or transported to the embryonic axis (emerging root and shoot) for the synthesis of protein and other nitrogenous compounds. This transport causes an increase in the amino acid and protein content of the axis. In many species, the nitrogen of the amino acids is converted to amido nitrogen of glutamine, which is the principal nitrogen containing compound exported from cotyledons and endosperm to the developing axis. Other interconversions of the amino acids, prior to their transport to the embryonic axis, also take place in the cotyledons or endosperms.

Some of the amino acids in embryonic axis may undergo transamination and produce keto acids. These keto acids are eventually metabolised through the TCA cycle to generate carbon skeletons and energy.

Protein Changes During Senescence: Each plant organ after attaining a definite age and maturity, reaches a stage of senescence. During senescence those deteriorative processes are initiated, which lead ultimately to the death of the organ. The metabolic activities fall rapidly and the reserves are translocated out to support the growth of younger organs.

During ontogeny of a leaf, the protein content is maximum in a fully developed and mature leaf. But the protein content declines during the onset of senescence. Other macromolecules, such as chlorophyll and DNA also decline during senescence. Yellowing of the leaf is the most visible symptom of senescence. However, decline in protein content starts well before any visible sign of yellowing is seen. Detachment of leaf accelerates the decline in protein and other macromolecules.

Decline in protein content during senescence appears to be both due to decreased protein synthesizing capacity and due to increased protein degradation in the leaves. Experiments with labelled amino acids have indicated that their incorporation into protein is inhibited or completely stopped during senescence and detachment of leaf. Increased protein degradation during senescence is apparent by increase in amino acids and amide content of the leaves and by the appearance or increase in the activity Proteases. Proteases from mature leaves have been isolated and characterised by many plant physiologists.

Protein Synthesis

Amino Acid Biosynthesis: Amino acids are the precursors of proteins. In plants, they are synthesized from inorganic nitrogen and organic acids. Inorganic nitrogen is taken in by the plants either in the form of nitrate or ammonium. Nitrate is reduced to ammonium, which is then incorporated into amino acids. Organic acids are derived from TCA cycle. Some of the basic mechanisms of amino acid biosynthesis are as follows:

Reductive Amination: Amino acid glutamate (glutamic acid) is synthesized by reductive amination of the keto acid, 2-oxoglutarate. It is the primary reaction in the incorporation of ammonium into amino acids. The reaction is catalysed by the enzyme glutamate dehydrogenase;

$$\text{2-oxoglutarate} + NH_3 \underset{NAD(P)H \quad NAD(P)}{\overset{\text{Glutamate dehydrogenase}}{\rightleftharpoons}} \text{Glutamate} + H_2O$$

Glutamic acid serves as a precursor for many other amino acids such as lysine, proline, arginine, etc. It also participates in transamination reactions.

As described earlier, an alternate mechanism of glutamate synthesis is through the amide glutamine, which itself is synthesized from glutamate and ammonia;

$$\text{Glutamate} + NH_3 \xrightarrow[ATP \quad ADP + Pi]{\text{Glutamine synthetase}} \text{Glutamine}$$

Glutamine in turn is able to aminate 2-oxoglutarate to produce glutamate;

$$\text{Glutamine} + \text{2–oxoglutarate} \xrightarrow{\text{Glutamate synthesis}} \text{2–glutamate}$$

In sum, one glutamate is used in the reaction and two glutamate molecules are produced giving a net production of one glutamate molecule. This pathway of glutamate synthesis is also called glutamine synthetase- glutamine oxoglutarate aminotransferase (GS-GOGAT) Pathway. Reductive amination of pyruvate to form alanine has also been observed in several micro-organisms;

$$\text{Pyruvate} + NH_3 \underset{\text{NADH} \quad \text{NAD}^+}{\overset{\text{Alanine dehydrogenase}}{\rightleftarrows}} \text{Alanine} + H_2O$$

In higher plants, however, alanine is synthesized by transamination reaction.

Transamination: Transamination is the most important reaction in the synthesis of plant amino acids. In this process, the amino group of a amino acid (donor amino acid is transferred to a keto acid, which is then converted to an amino acid. This type of reaction is catalysed by enzymes called transaminases or aininotransferases.

Aspartate amino transferase (glutamate oxaloacetate transaminase) catalyses the synthesis of aspartate;

$$\text{Glutamate} + \text{oxaloacetate} \rightleftarrows \text{2-oxoglutarate} + \text{aspartate}$$

Transamination reactions are also involved in the synthesis of glycine, alanine, isoleucine, leucine, valine, serine and the aromatic amino acids. Transaminases require pyridoxal phosphate as coenzyme.

The transamination is also important in the metabolism of non-protein amino acids, in particular ornithine and gama-amino butyrate.

Mechanism of Provein Synthesis: The molecular mechanism of protein synthesis, starting from amino acids, is almost similar in plants, animals and micro-organisms. In plants, most of the proteins are synthesized in the cytosol, although a few are synthesized in chloroplasts and mitochondria.

The amino acids are transported from the cytosolic pool to the specialized cell organelles called ribosomes, by transfer RNA (t-RNA), where they join sequentially through peptide linkages to form a polypeptide. The sequence of amino acids in the polypeptide is determined by the sequence of nucleotides in DNA (gene). First, the DNA is transcribed to a messenger RNA (m-RNA) molecule, which has a nucleotide sequence complimentary to that of DNA. The m-RNA then moves from nucleus to the cytosol, where its nucleotide sequence is translated into the amino acid sequence of the proteins. The nucleic acid directed protein synthesis is represented by following line diagram, and is often considered as 'Central dogma' of molecular biology;

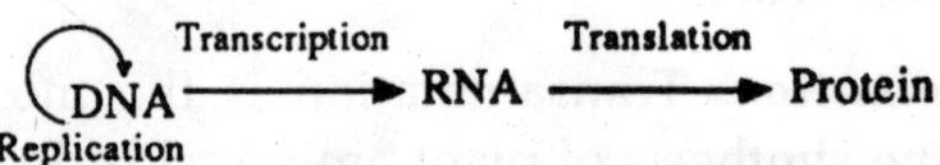

The translation process, *i.e.*, formation of polypeptide by various amino acids as directed by nucleotide sequence of m-RNA has been covered in this section. The process involves following steps:

Activation and Transfer of Amino Acids: The first step in protein synthesis is the activation of individual amino acids and their transfer to the ribosomes by t-RNA. The process is catalysed by the enzyme amino acyl t-RNA synthetase, which is specific for each amino acid and each t-RNA. In bacteria, a single synthetase for each amino acid has been found, while in eukaryotes multiple synthetases for the same amino acid are found in the cytosol. Each synthetase has at least two binding sites; one for the amino acid and other for its specific t-RNA. In the reaction process, which is accomplished in two steps, the amino acid is first activated by ATP in the presence of amino acyl t-RNA synthetase. Amino acid forms a covalent link with AMP (product is called amino acyl adenylate) and pyrophosphate is liberated;

$$\text{Amino acid} + \text{ATP} \xrightarrow{\text{Amino acyl t–RNA synthetase}} \text{Amino acyl adenylate} + \text{PPi}$$

In the formation of amino acyl adenylate, an anhydride link between carboxylic group of amino acid and phosphate group of AMP is established. The pyrophosphate (PPi) liberated can be converted to two molecules of inorganic phosphate by the enzyme pyrophosphatase;

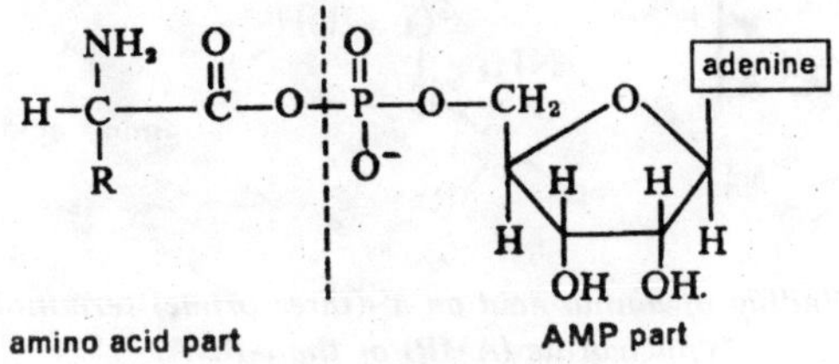

$$\underbrace{\mathrm{H{-}C(NH_2)(R){-}C({=}O){-}O}}_{\text{amino acid part}}\;\Big|\;\underbrace{\mathrm{{-}P({=}O)(O^-){-}O{-}CH_2{-}ribose(OH, OH){-}adenine}}_{\text{AMP part}}$$

amino acid part AMP part

$$\text{PPi} \xrightarrow{\text{Pyrophosphatase}} 2\ \text{Pi}$$

In the next reaction, the complex amino acyl adenylate, which remains bound to the enzyme amino acyl t-RNA synthetase hits the specific t-RNA; amino acid is transferred to t-RNA and AMP is liberated:

$$\text{Amino acyl adenylate} + \text{t-RNA} \xrightarrow{\text{Amino acyl t–RNA synthetase}} \text{Amino acyl t-RNA} + \text{AMP}$$

The amino acid is accepted by the terminal (5') adenosyl nucleotide of the t-RNA. This end of t-RNA is also called as amino acid acceptor end. The amino acid reacts with either 2' hydroxyl or 3' hydroxyl group of the ribose moiety of terminal adenine nucleotide. The amino acid is transported to ribosomes in this form.

Formation of Initiation Complex: Initiation of protein synthesis requires amino acyl t-RNA complex (as formed in the first step), m-RNA with initiating codons (AUG), 40 S and 60 S sub units (30 S and 50 S in prokaryotes) of ribosomes, ATP, GTP and certain specific proteins called initiation factors (IF). In plants and other eukaryotes, the first amino acid to be incorporated in proteins is methionine while in prokaryotes it is formyl methionine. Different components of protein synthesis initiation, join in a definite order to form an initiation complex. The sequence is:

Binding of amino acid on 3′ (three prime) terminal nucleotide (AMP) of the t-RNA.

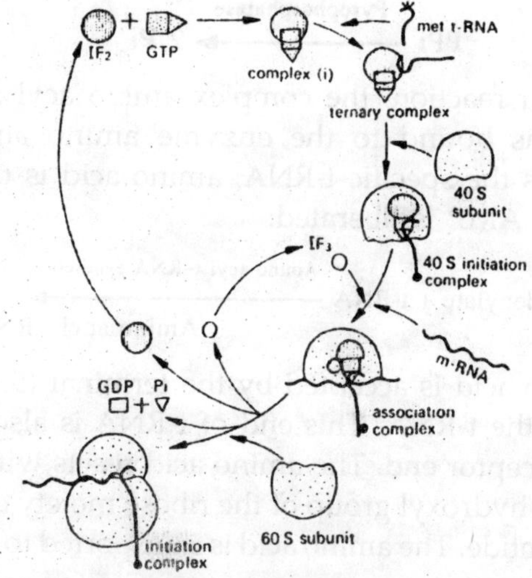

Formation of initiation complex

(1) The energy rich nucleotide GTP-binds with initiation factor 2 (IF2), increasing the affinity of IF2 for methionine t-RNA (met t-RNA);

$$\text{GTP} + \text{IF2} \rightarrow \underset{\text{complex (i)}}{\text{GTP-IF2}}$$

(2) The t-RNA charged with methionine reacts with complex (i) to form a ternary complex;

GTP-IF2 + met t-RNA → met t-RNA-IF2-GTP

Ternary complex

(3) The ternary complex then associates with 40 S (smaller) sub unit of the ribosome. The complex thus formed is referred to as 40 S initiation complex;

met t-RNA-IF2-GTP + 40 S → 40 S-met t-RNA-IF2-GTP

40 S initiation complex

(4) Initiation factor 3 (IF3) catalyses the formation of another complex between 40 S initiation complex and m-RNA containing initiating codon. The process is assisted by IF1, IF4A, IF4B and by high energy molecule ATP. The m-RNA binds to the 40 S ribosomal unit at its 5' end;

40 S–met t-RNA–IF2–GTP + m-RNA —(IF3, IF1, IF4A, IF4B; ATP → ADP + Pi)→
m-RNA–40 S–met t-RNA–IF2–GTP
Association complex

(5) Association complex moves towards the 3' end of the m-RNA in search of an initiating codon. Then 60 S sub unit of the ribosome attaches to this complex. The initiation factor 5 (IF5) assists in this attachment. This factor helps in releasing IF2 and IF3 from association complex. 60 S sub unit does not attach to this complex unless IF2 and IF3 are released. The attachment itself if catalysed by factor IF4C, and GTP of the initiation complex. GTP of the initiation complex is hydrolysed to provide energy for it;

m-RNA–40 S–met t-RNA–IF2–GTP + 60 S —(IF5)→
m-RNA–80 S–met t-RNA + IF2 + GDP + Pi

This complex is now ready for the formation of polypeptide chain in the subsequent steps.

Polypeptide Chain Formation: After the formation of initiation complex, polypeptide is formed by regular addition of amino acids, which are activated and then brought to the ribosomes by their respective t-RNAs.

There are two sites on the ribosomes; the peptidyl site or 'P' site and the amino acyl site or 'A' site. The initiating amino acid-mRNA-tRNA has to be on 'P' site before the next amino acid can come and bind on 'A' site. Elongation factor Ts.Tu (discovered in bacteria) assists in this binding. GTP is also required for the binding. The series of reactions involved in this binding are;

$$\text{TsTu} + \text{GTP} \rightarrow \text{TuGTP} + \text{Ts}$$

$$\text{TuGTP} + \text{aa-tRNA} \rightarrow \text{aa-tRNA-TuGTP}$$

This ternary complex (aa-tRNA-TuGTP) is on 'A' site of the ribosome. After binding to ribosome, GTP is hydrolysed;

$$\text{aa-tRNA-TuGTP} + \text{ribosome} \rightarrow \text{aa-tRNA-ribosome} + \text{TuGDP} + \text{Pi}$$

and TuTs is regenerated to start another cycle for the binding of the amino acid charged on transfer RNA (aa-tRNA);

$$\text{TuGDP} + \text{Ts} \rightarrow \text{Tu-Ts} + \text{GDP}$$

Later on, formation of peptide bond between the free carboxylic group of the amino acid on peptidyl RNA at the 'P' site and free amino group of the amino acid on amino acyl t-RNA at 'A' site, takes place. This reaction is catalysed by the enzyme peptidyl transferase. After the formation of peptide bond, the t-RNA present at 'P' site is deacylated (uncharged) and liberated from the site. Transfer RNA present at 'A' site is then translocated to 'P' site. This process requires GTP and elongation factor G. GTP is hydrolysed to GDP and Pi in this process, to liberate energy, by the factor G. Each time a peptidyl group is translocated, a GTP molecule is hydrolysed.

Termination of Polypeptide Chain: Termination of polypeptide chain is signalled by any of the three termination codons in m-RNA, UAG, UAA or UGA. Normally no t-RNA contains an anticodon which is complimentary to these codons. When a termination codon appears at 'A' site of the ribosome and completed polypeptide occupies the 'P' site, a release factor RF-1 comes to play its role.

The release factor alters the activity of the peptidyl transferase causing it to hydrolyse the amino acyl ester bond holding the C-terminal amino acid residue to the t-RNA. The same or other release factor in association with GTP releases the final polypeptide. The released polypeptide is then modified in various ways to achieve its desired role in the cell. At this point, the ribosome dissociates into 40 S and 60 S sub units, which enter the cytosol to enter another cycle of protein synthesis.

It should be realised that protein synthesis involves the simultaneous action of many ribosomes on a single m-RNA molecule. This complex between a m-RNA and several ribosomes is known as a *polyribosome* or *Polysome.* Several such polysomes can be seen in an actively protein synthesizing cell.

Some Unusual Plant Proteins: Most of the proteins in plants function either as enzymes catalysing various biochemical reactions or as structural components of cell and cell organelles. Some proteins also function in the motility and transport of some lower forms of plants. However, a few specialized proteins are also known to exist in plants and also in other living systems, which perform some unusual functions after appropriate physico-chemical modifications. Some of these are described in following paragraphs.

Lectins: The term 'lectin' has been used for those proteins which have agglutinizing properties. The property is apparently because they have specific sites, where carbohydrates can bind. According to N. Sharon and H. Lis (1972) lectins are defined as "sugar binding protein or glycoprotein of non-immune origin which agglutinates cells and/or precipitates glycoconjugates".

Lectins have been found in a variety of species including non-flowering and lower plants. The seeds of legumes are particularly rich in lectins, and many of these lectins have been characterised extensively. In many legumes, about 10% of the soluble seed protein is lectin, where they are localised inside the protein bodies of the cotyledon cells. Lectins from various sources are similar in many characteristics, although they vary as well in several aspects. Properties of some representative lectins are listed in table.

Characteristics of Some Lectins

Lectin	*Distribution and source*	*M.W. kDa.*	*Subunits*	*Composition*
Concanavalin A	Diocleae tribe of leguminosae (seeds)	104	Tetramer of identical subunits	Rich in asparagine, serine, little methionine, no cysteine, no carbohydrates
Favin	Vicieae Tribe of leguminosae (seeds)	53	Tetramer of 2 chains	Rich in asparagine and threonine, no methionine or cysteine, 3% carbohydrate
Soybean lectin	Phaseoleae tribe of leguminosae (seeds)	120	Tetramer of identical subunits	Rich in asparagine, threonine and serine; little methionine, no cysteine, 7% carbohydrate
Potato lectin	Solanaceae (tubers)	50 or 100	Monomers or dimers	Rich in hydroxy-proline, cysteine, serine and glycine; no phenyl alanine, 50% carbohydrate
Wheat germ agglutin	Graminaceae (embryos)	36	Dimer of identical subunits	Rich in cysteine and glycine; No carbohydrate

The exact role of lectins in plants is unknown. However, following hypotheses have been proposed about their function :

(1) Lectins may function in seed maturation, dormancy and germination. In this context they may function as storage proteins, or in the packaging and mobilization of storage materials.

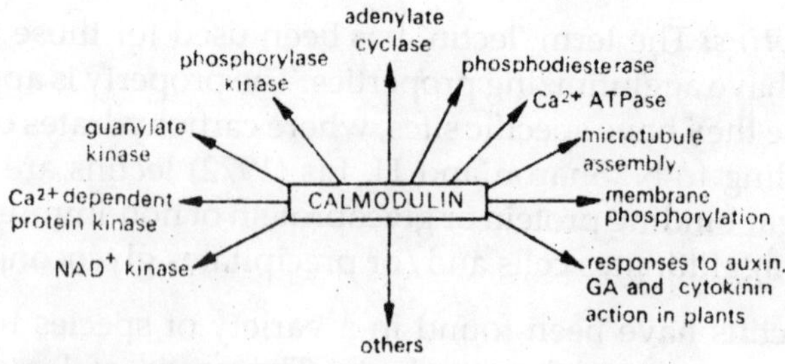

Calmodulin regulated enzymes and cellular processes.

(2) Lectins may protect plants against bacterial, fungal and viral pathogens during seed germination and early seedling growth.

(3) Lectins may be involved in the symbiotic association of *Rhizobium* with legume roots.

(4) They behave as enzymes, for some specific carbohydrates.

Calmodulins

Calmodulins are proteins which reversibly bind with calcium and affect many cellular and biochemical processes. Its presence has been reported in both plants and animals. In plants, comparatively higher levels of calmodulin are found in root meristem. The amount increases during seed germination and seedling growth. As will be evident from the account on mechanism of action of plant hormones, calmodulins play an important role in plant hormone action because of their calcium binding property. It should be realised that only proteins have the affinity to bind with Ca^{2+} (but not nucleic acids, carbohydrates or phospholipids) because they offer carboxylate groups that bind Ca^{2+}.

Calmodulins are acidic in nature, because they contain many acidic amino acid residues. Most of them lack cysteine, hydroxyproline and tryptophan. The presence of acidic amino acids provides carboxylate groups for the reversible binding of Ca^{2+}. Lack of cysteine and hydroxyproline provides a tertiary structure highly flexible to interact with the various calmodulin regulated proteins. Calmodulins are relatively thermostable.

Calmodulin in association with Ca^{2+} induces a variety of cellular and enzymic processes. These enzymes control not only important steps in cellular metabolism, but also the metabolism of certain key regulators. For example, in animal systems calmodulin governs both the synthesis and degradation of cyclic AMP. The mode of action of calmodulin on enzyme system was first established with the phosphodiesterase system.

Calmodulin itself is not active, it is activated after binding with Ca^{2+};

$$(\text{Calmodulin})\ \text{inactive} + Ca^{2+} \rightleftarrows (\text{Calmodulin} . Ca^{2+})\ \text{active}$$

Calcium bound calmodulin assumes a new helical structure which is active and binds with the enzyme. The enzyme in turn is activated to carry on the desired biochemical response.

$$(\text{Calmodulin.}Ca^{2+})_{active} + \text{less active enzyme} \rightleftarrows (\text{Enzyme.Calmodulin.}Ca^{2+})_{active}$$

Arabinogalactan Proteins

Arabinogalactan proteins (AGP) are found in many higher plants and in many of their secretions. They are specially abundant in the species from orders Myrtales, Rutales, Rosales, Fabales and Proteales of angiosperms and Cycadales, Coniferae and Gnetales of gymnosperms. Gums, secreting out of many plants are particularly rich in AGP.

Chemically, AGP are glycoproteins or proteoglycans (protein + carbohydrates) having molecular weight in the range of 10^4 to 10^6. They contain a high proportion of carbohydrates, in which arabinose and galactose are predominant monosaccharides. Other monosaccharides which may be present are: L-rhamnopyranose, D-mannopyranose, D-xylopyranose, D-glucopyranose, D-glucuronic acid and D-galacturonic acid. The protein content of AGP is small, usually between 2 and 10%, but may be as high as 59% in *Acacia hebeclada* gum. The proteins are specially rich in hydroxyproline, and also in serine, alanine and glycine. In AGP, the hydroxyproline of the protein component, is linked through a glycosidic linkage to the (J-D-galactopyranose residue of the carbohydrate component. Possible functions of AGP are as follows:

(1) As AGP are present in plasma membrane they may be receptors of external signal such as the low m.w. β-glucan elicitors of phytoalexin response. Phytoalexins

are antimicrobial compounds of low m.w. that are synthesized in plants after infection with micro-organisms.

(2) AGP may play a role in wound healing, frost hardiness and drought resistance, as they have high capacity for water holding.

(3) Like animal glycosaamingIycans, the AGP of plants may be involved in cell wall adhesion.

Phytochelatins

Phytochelatins were previously called as metallothioneins as in animals. However, it has been shown recently that phytochelatins are different from metallothioneins in many respects. They are small proteins involved in detoxifying certain heavy metals such as cadmium, lead, copper, zinc, etc. They are induced in a whole range of plants, including algae, upon exposure to wide range of heavy metals. Roots contain higher amounts of phytochelatins than other parts of the higher plants.

Most of the phytochelatins are rich in glutamate and cysteine; there are 2 to 11 (Glu-Cys) sequences. Glutamate is in the form of y-glutamyl residue. Because of this, it is considered that phytochelatins are not primary gene products, but they are probably synthesized by certain enzymes which are induced by heavy metals. In most cases, C-terminal amino acid is glycine, but for in order Fabales, where it is β -alanine. The heavy metals bind to phytochelatins via thiolate coordination through cysteine residue.

The m.w. of phytochelatins isolated from algae is in the range of 8 to l3kDa. In other plants, it exists as a smaller oligomeric protein of 2 to 10 kDa.

About the physiological role of phytochelatins, they are believed to be the proteins conferring resistance to plants to heavy metals. They are induced by heavy metals and are normally absent in those plants which grow on uncontaminated soils. The fact that root tissues contain a much higher concentration of heavy metals as well as of phytochelatins than the leaf tissues, points to the fact that metals are immobilized to a far greater extent at the site of metal uptake.

P-proteins

The term P-protein was introduced by K. Esau and J. Cronshaw (1967) for certain proteins found in the sieve tubes of the phloem. These proteins, have been detected in sieve elements of all dicotyledonous and a few monocotyledonous species. They are seen as a discrete aggregates of molecules. Sometimes, they also appear as loose mass of protein. They may be tubular and fibrilar, granular or even crystalloid in appearance. Phloem exudates can be collected from the plants and the structure of P-proteins can be examined under the microscope.

The P-proteins are apparently made up of several sub units, the m.w. of each unit ranging from 10 to 13kDa. They are rich in lysine residues.

Although the exact role of P-proteins is not known, they seem to play an important role in the structural organization and function of sieve tubes. Some suggested roles are as follows:

(1) They may participate in phloem transport as some form of contractile protein.

(2) They may be involved in sealing the injured sieve elements.

(3) Possibly they are involved in some kind of recognition system including host-parasite relationship.

Electron Transfer Proteins

Several proteins participate in the mitochondrial and photosynthetic electron transfers. These proteins exist in oxidized as well as reduced forms. The two forms are inter-convertible. The reduced forms of these proteins are strong reducing agents. Some of these proteins are described briefly over here.

Cytochrome C: Cytochrome C has been the most extensively studied electron transfer protein, involved in electron transport in almost all eukaryotes. In general, plant cytochrome C are similar to those from all other eukaryotic sources. The protein consists of a single polypeptide chain of about 103-113 amino acids, without any disulfide bridge. A heme group is attached to the polypeptide chain by thioester linkage to two cysteine residues.

Cytochrome C_6: Cytochrome C_6 (or C_{552}) is a soluble cytochrome found in certain algae and is associated with photosynthetic electron transfer. It has a m.w. of about 10kDa.Heme group is attached to the polypeptide through two cysteine residues present at 14- and 17- residues.

Cytochrome f: Cytochrome f is present in higher plants and is tightly lipid bound in most cases. It has a m.w. of about 33kDa and is structurally similar to cytochrome C_6.

Plastocyanin: Plastocyanin occurs in many algae and higher plants and functions as an electron transfer protein in photosynthesis. It has a m.w. of about 11 kDa and contains one copper atom per molecule.

Ferredoxin: Ferredoxin are small non-heme proteins with a m.w. of about around 11 kDa. It acts in electron transfer in a variety of reactions, including photosynthesis and nitrogen fixation. In anaerobic bacteria, it is involved in pyruvate metabolism. It contains about 93 to 98 amino acids. There are four cysteine molecules at 39-, 44-, 47- and 77- residues. The iron atoms are bound to the protein through these cysteine residues.

Histones

Histones are small basic proteins found in association with DNA. Their mass is almost equal to that of DNA. These proteins are rich in basic amino acids: arginine and lysine. There are five main types of histones in plants;

H_1 type — very lysine rich

H_2b type — lysine rich

H_2a type — arginine-lysine rich

H_3 type — arginine rich

H_4 type — glycine arginine rich

The chemical structure of plant histones is almost similar to animal histones.

8

Physical Development

Generally, the insects develop from the eggs, which are formed from primordial cells in the ovaries of females. Eggs hatch outside the female body and such females are known as oviparous but in certain cases they do so within the body, such females being termed ovoviviparous. Still there is another group of females, which give rise directly immature stages known as viviparous. Even some females are able to reproduce normally without mating with the males, this phenomenon is termed as parthenogenesis.

One of the most characteristic feature of insects is that they generally hatch in a condition which is morphologically different from the adult. In order to reach the adult stage they have to pass through a number of changes which are collectively termed as metamorphosis. The growth of an insect within the eggs is called as the embryonic development and the changes in the insect development after hatching is termed as post embryonic development. Every insect during its growth casts its skin and this process is known as moulting or ecdysis. The skins casted by the insect are known as exuviae. The interval between the moulting is known as stage or stadium and the form assumed by the insect during a particular stadium is termed as instar. The final instar is a full developed form known as adult or imago.

Types of Metamorphosis

According to Essig, the metamorphosis may be classified into two main groups:

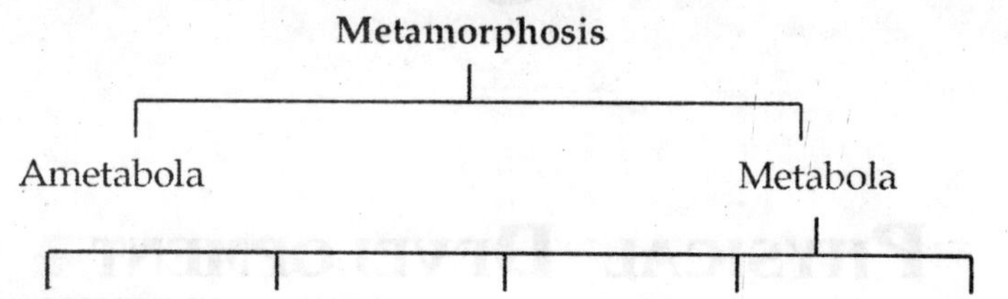

Ametamorphosis

This is also known as Ametabola in which no distinct external changes are observed in adults and immature stages (nymphs), except in size and colour. This type of metamorphosis is found in insects belonging to the Protura, Thysanura, Diplura and Collembola of Aptergota group. According to Imms certain apterous forms in the phasmida, Anoplura and Isoptera may also be placed in this group.

Ametabola: In primitive wingless insects or Apterygota only very slight changes are brought about during post-embryonic development at each moult and moulting continues even after the stage is reached. Such insects which emerge from the egg looking like the adult in appearance and which do not have a metamorphosis in strict sense.

Metabola: In this type of metamorphosis there are distinct changes during growth and development. This type may be divided into the following groups:

(i) *Paurometabola: This* is also known as simple, direct or gradual metamorphosis in which the immature stages resemble the adults except in size. The immature stage is called as nymph in which the wings and genital organs are not developed. This type of metamorphosis is met in most of the exopterygota specially in order Orthoptera, Hemiptera and Isoptera.

(ii) *Hemimetabola:* This is also an incomplete type of metamorphosis in which accessory organs known as gills are found in nymphal stages. The nymphs of such insects are termed as naiads which live in water while adults are aerial. The representative of this groups belong to the Order Odonata and Ephemeroptera.

(iii) *Holometabola:* This type of metamorphosis is also termed as complete, indirect and complex type of metamorphosis in which four stages of insect are found during it's development. These are Egg-Larva-Pupa and Adult. The larvae and pupae are quite different from the adults and morpholgically larvae may assume several different forms.

Types of Larvae

Oligopod: These are characterized by the presence of more or less well developed thoracic legs and the absence of abdominal appendages. However, in certain cases a pair of cerci or caudal processes are present. They may be further divided into two groups:

(i) *Campodeiform Larvae:* These are elongated with long or short sickle-like mandibles, and often well developed antennae and cerci. They have no compound eyes or ocelli. They are represented by members of the order Ephemeroptera, Odonata, Neuroptera, Plecoptera and Strepsiptera.

(ii) *Scarabaeidform Larvae:* They are stout, sub-cylindrical and C-shaped. They have short thoracic legs, soft fleshy body without abdominal legs. They are found in order Coleoptera.

Protopod: These larvae are found in some parasitic forms of order Hymenoptera, which emerge early in their embryonic development and have the internal systems poorly developed.

***Polypod*:** They have well defined segmented body with small thoracic legs and a number of abdominal pro-legs. The tracheal system well developed provided with a number of spiracles. The typicals of this type are seen in eruciform larvae of order Lepidoptera.

Apodous: They are degenerate type of larvae which are legless, robust, C-shaped or spindle shaped, with or without well developed head. They may be subdivided into three types on the basis of the degree of the head development and their shape:

(i) *Curculionoid:* These are robust C-shaped with well developed head and also known as Eucephalus. It is common in most Coleoptera and some Diptera.

(ii) *Muscoid:* These are spindle shaped or cylindrical and truncate. They have no obvious head capsule and therefore, sometimes also termed as Acephalus. This type of larvae are known as maggots which are commonly found in the flies of the order Diptera.

(iii) *Apoid:* These are robust with well developed head, cared for by daily feeding in provisioned brood cells. This is common in ants, bees, wasps and certain parasitic forms.

Types of Pupae: This instar differs from the larvae as it is more or less quescents which does not feed and it is the site of more or less profound internal changes which transform the larval organization into that of the adult. Two types of pupae are recognized:

1. *Decticous Pupa:* This type of pupa is provided with functional mandibles which are used by the pharate adult to break out of the cocoon or pupal cell before the adult emergence. It is found in Neuroptera, Meeoptera, Trichoptera and a few primitive Lepidoptera.
2. *Adecticous Pupa:* This type of pupa is found in other endoptcrygotes, has no functional mandibles. It may

escape from the cell with the aid of various spines or other special processes. It may further be divided into two:

(a) *Adecticous Exarate Type:* In this case the appendages projecting freely from the body, *e.g.*, Coleoptera, Hymenoptera.

(b) *Adecticous Obtect Type:* In this case the appendages like wings, leg, antennae and mouth parts firmly soldered down to the body, *e.g.*, In most Lepidoptera.

(iv) *Hypermetabola:* This is more complex type of metamorphosis in which there are several types of larvae differing from each other after each instar. For example, a minute active first instar, a more or less robust and sluggish second instar, and a similar but apodous third instar. Such insects belong to order Neuroptera, Coleoptera (Meloidae, Carabidae etc.) Strepsiptera and parasitic Diptera.

(v) *Poly-metabola:* In some insects from a single egg more than one larvae have been seen to hatch out and this process is termed as poly embryony and the metamorphosis is known as poly metabola. This type of metamorphosis is found in the insects belonging to the families Itonididae and Chalcididae.

The Metamorphosis

The grasshoppers undergo simple or incomplete metamorphosis. The female grasshopper selects the loose textured, easily penetrable soil for egg laying. The gravid female, when ready for oviposition, tries its abdomen on the ground in search of suitable place by scraping the soil surface with its ovipositor. After searching the suitable spot, the female bores the soil by rotatory movements of the abdomen and makes the egg chamber which may go up to 6 inch deep in the soil. When egg chamber is formed the female secrets the gelatinous fluid or frothy which helps in cementing the walls of chamber as the fluid gets hardened after drying. The eggs are laid one by one arranged longitudinally in a spiral fashion; after which the female again secrete the fluid which closes the mouth of chamber. Each egg chamber contains 60-100 eggs. Later, the female covers the egg chamber by loose soil with the help of its hind legs. This structure is termed as egg pod. The female deposits two or three egg pods in her life time.

Egg: The egg of grasshopper closely resembles with rice grain. It is elongated, curved, thick in the middle and tapering towards each end which are blunt and rounded. There are two layers known as the chorion and the vitelline which is thin and lined inside chorion. At the time of hatching it changes into the brown colour under which the young one can be seen moving its abdomen and hind femora. Due to the movement of young one, the egg ruptures with the help of two cervical ampullae and nymph comes out of it. The newly hatched nymph starts feeding after 24 hours of emergence and passes through 5 or 6 moults before becoming an adult.

First Instar Nymph: It is similar to the adult in general body structure but smaller and wingless. The mouth parts are well developed and adapted for chewing and cutting type. The eyes are oval, black and without stripes. The head is conical and projected forward having eight segmented antennae. The pronotum is saddle shaped and broader than meso- and meta-thorax. The fore and middle pairs of legs are small, while the hind pair is stout, long and adapted for jumping type. Tarsi are three segmented provided with a pair of claws in the terminal segment and a median pad — the arolium.

The wings traces are not present. In both sexes ten terga are distinct of which the last two are partially fused but the number of apparent sterna in the two sexes are different. The male has ten sternal segments including the first one fused with the meta thoracic sternal plate. The ninth sternum is large, ventrally depressed and deeply notched at the posterior margin which is considered as the rudiment of the subgenital plate or gonocoxite.

The podical plates are found above gonocoxite which are paired triangular outgrowths. In case of female the ninth sternum is cleft into two triangular pieces which are the rudiments of dorsal valves of ovipositor. In the posterior margin of eighth sternum there is a tiny pair of protruberances which are the rudiments of the ventral valves. The nymph after attaining the maximum possible growth sheds off its skin and called as IInd instar nymph.

Second Instar Nymph: During moulting the cervical ampullae exert pressure to burst the cuticle lengthwise and through a slit thus made, emerges out second instar having a new soft skin. It is similar in colour and general appearance to the first stage nymph except the size. The antennae are nine segmented and compound eyes develop a transverse whitish stripe on each. The meso- and meta-thorax become some what extended outward but there appears no indication of wing pads. The subgenital plate of ninth sternum in male increases in size and the notch flattens out gradually. The ventral and dorsal valves of the ovipositor grow longer and extended backwards covering the part of the ninth segment and the anterior half of the podical plates, respectively.

Third Instar Nymph: It is larger in size than the second instar, colour becomes deeper and the white stripe in these gets broader. The antennae are 10 to 12 segmented. On the lateral side of meso- and meta-thorax, the distinct thin triangular out growths develop which are the rudiments of the wings. In males the posterior margin of ninth sternum becomes rounded which forms gonocoxite.

Fourth Instar Nymph: The colour of this instar is deeper than the third instar and also bigger in size. The antennae are 13 to 14 segmented. The wing pads of hind pair overlap the fore pair and reach upto the first abdominal segment. The wing pads are transparent, light green in colour and rudiments of the veins appear. The subgenital plate in male becomes conical and spatula like in appearance and compresses the podical plates. The ventral valves in female extend further and become narrow and pointed.

Fifth Instar Nymph: After fourth moulting comes this instar which is bigger in size than the former and provided with fifteen segmented longer antennae. The thorax is entirely covered with pronotum. The wing pads reach upto the third abdominal segment and veins become comparatively more prominent. The subgenital plate reaches the supra anal plate. In female the ventral valves become thicker and reach the middle of the dorsal valves. The egg size becomes bigger and distinct. After this moult it generally changes into adult but sometimes there is sixth instar also.

Sixth Instar Nymph: This is very similar to adult except that wings are slightly smaller. The transverse stripes of the eyes disappear altogether. The subgenital plate in male becomes a pouch like organ into which the aedeagus is lodged. In female, the egg guide remains pushed in between the two ventral valves and the dorsal valves become very thick and extend beyond the podical plate.

Adult: In eyes, the dot or white stripe is completely absent. Antennae are 16 segmented and hind wings are overlapped by the fore wings and cover the abdomen. Cervical ampullae disappears. In male, the subgenital plate is equal to supranal plate. The ventral and dorsal valves of the ovipositor become highly chitinous and the lateral lobes of ventral valves become very large and appear as distinct sclerite.

Metamorphosis in Butterfly

The butterfly undergoes to complete metamorphosis *i.e.,* it has all the four developmental stages namely egg-larva-pupa and adult. Generally, all the butterflies pass through these stages of metamorphosis more or less in the same manner. Therefore, the example of a common butterfly *(Papilio demoleus)* has been described. The female butterfly oviposits on the lower surface of the leaves, singly or in batches close to mid rib.

Egg: Each egg is conical or oval in shape measuring about 1.2 mm in length and 0.6 mm in breadth. The colour of egg is olive green or dark green with a number of longitudinal ridges with a few horizontal lines. The egg hatches within a week and this period is known as incubation period or hatching period. The egg shell is generally eaten up by the emerging larva.

First Instar Larva: The worm just after hatching is called as first instar larva which is about 3 mm long. The body consists of head, three thoracic and nine abdominal segments. The head is hypognathous, black and bigger than other segments. The mouth parts of the larva are biting and chewing type and it starts feeding immediately after hatching. Each thoracic segment carries pair of 3 jointed legs known as thoracic legs while abdominal legs called prolegs are present from each of the third to sixth and last abdominal segments. Each pair of abdominal legs bear crochets on the distal end which are used for catching the leaf during the feeding period.

The meso-thorax bears a pair of tubercles. Ten pairs of spiracles are found of which two are situated on meso- and meta-thorax and the rest on the abdominal segments. The larva grows in size and when the cuticle is not able to accommodate the developing body, it bursts away and second instar larva comes out.

Second Instar Larva: The larva is nearly twice as big as the first instar and on its blackish head, a whitish clypeus is visible. The transverse lines over the body become more deeper. The legs, ventral region of the abdomen and spiracles become black. A pair of setae arises on the meso thoracic tubercles which are erected just like antennae. The similar upward projected setae are also found on the 8th abdominal segment. After some time the larva moults and changes into third instar.

Third Instar Larva: It is about 1.5 times bigger than the IInd instar larva. The genae around the clypeus also get whiten. Besides, more of prominent transverse lines appear over the body and spiracles become oval in shape surrounded by black rings. The setae of the meso thorax and eighth abdominal segments are lengthened and show movement when larva is disturbed.

Fourth Instar Larva: One more whitish line develop in the centre of the head and forming the total three. It is about twice as long as third instar. The setae grow in size and thickness and more transverse lines appear on the body. At the end moulting takes place and fifth instar larva emerges out.

Fifth Instar Larva: The larva of this instar is 4-5 cm in length with light green or bluish colour having yellow linings and several black transverse linings. The setae of prothorax become yellow at the base and blackish towards the distal end. The spiracles become oval in shape and prominent. One more spot develops on the head. When ready for pupation it forms a black cord known as cremaster by the silken threads produced from the mouth. The larva then hangs upside down attaching its distal end of the abdomen to the cremaster. The colour of the larva gradually faints and the body shrinks. The legs disappear and it changes into pupa. This is transformation from larva to pupa. The transformation from larva to pupa takes about 8-9 hours.

Pupa

Newly formed pupa is of light green in colour and oval or conical in shape also called as chrysalis. Pupa is covered by puparium and has golden spots on the distal end. The silvery white spots appear on the various parts of the head and thorax. The spiracles are also visible on either side of the puparium. With the advancement of development, wings, eyes and antennae are visible through the pupal skin. It takes about 7-10 days to develop in the adult butterfly known as imago. At the time of emergence it changes into yellowish brown in colour.

Imago, after hatching it rests for sometime and shrinks and spreads its wings then flies away.

The whole life cycle is completed within 3 weeks in summer and 4 weeks in winter depending upon the climatic conditions.

GRASSHOPPERS

Among vertebrates the nervous system is dorsal while the circulatory system is ventral in position. Unlike these animals the condition is reverse in case of insects, *i.e.*, the circulatory is dorsal where as the nervous system is ventral and the alimentary canal is situated in between these two systems. The reproductive organs are located on either side of the alimentary canal in posterior region of the abdomen. The detailed description of the various systems is given under the different appropriate heads.

Various System

It includes the organs of ingestion (alimentary canal and its associated glands) and the physiology of digestion. The organs of ingestion are located in the head and are meant for the intake of food. They constitute the mouth parts which have already been described earlier.

The preoral cavity is enclosed by the mouth parts and is divided into two parts by the hypopharynx, the anterior region in which the alimentary canal opens is termed as cibarium and in which the salivary duct opens is known as salivarium. In the sucking insects the cibarium is modified into a sucking pump while salivarium serves as the salivary syringe. In larval Lepidoptera the silk-press is also a modification of salivarium.

Alimentary Canal: The alimentary canal of grass hopper is a simple, hollow and tubular in structure which runs from the buccal cavity to anus. It is distinctly divided into the following three primary regions:

1. Foregut or stomodaeum.
2. Mid gut or mesenteron or ventriculus.
3. Hind gut or proctodaeum.

Foregut or Stomodaeum: It constitutes the anterior region of the alimentary canal which is primarily an organ of ingestion and shows as a site for storing food. It consists of the following parts:

(i) *Pre-oral Food Cavity:* It has been described previously and indeed it is not a part of alimentary canal.

(ii) *Pharynx:* It is situated in between the pre-oral cavity and the oesophagous and is provided by the dilateral muscles. These muscles are highly developed in those insects in which pharynx helps in forming the sucking pump.

(iii) *Oesophagous:* It is a simple straight tube which runs from the posterior region of the head to thorax and joins with the crop.

(iv) *Crop:* It is simple bag like structure and serves as a storage reservoir for the food. Apparently it is a dilated portion of the oesophagous but differs histologically by the presence of sclerotized ridges which are arranged transversely in the crop. Since it serves as a reservoir for food hence its walls are thin and the muscles are poorly developed.

(e) *Gizzard:* It is situated in the posterior region of the crop which can not be apparently distinguished from crop but differs internally by having the longitudinal folds into the lumen in which cuticular teeth are attached. Its posterior part is concentric in the internal layer of six 'V shaped processes are attached which form the cardiac valve with the folds of gizzard. Its major function is to regulate the passage of food into the mid gut.

Histologically, the following layers may be distinguished in the walls of the stomodaeum:

(1) *Intima:* The inner most layer of chitin found in continuation of body cuticle.

(2) *Epithelial Layer:* It is a thin layer secreting the intima.

(3) Basement Membrane: Bounding the outer most surface of the epithelium.

(4) *Longitudinal Muscles:* These muscles are less developed than circulatory muscles.

(5) *Circulatory Muscles:* These are well developed.

(6) *Peritoneal Membrane:* It is often difficult to detect and consists of apparently structureless connective tissue.

Mid Gut or Mesenteron: It is relatively a short tube or elongated sac with uniform diameter extends from hepatic caecae or cardiac valve to Malpighian tubes or pyloric valve. Histologically, the inner wall of mesenteron or stomach is not made up of chitin, but consists of following layers:

(i) Peritrophic membrane

(ii) *Enteric epithelium*

(iii) Basement membrane

(iv) *Circular muscles*

(v) Longitudinal muscles

(vi) *Peritoneal membrane*

The enteric epithelium is made up of three types of cells: *(i)* The columnar cells which secret the enzymes and absorb the digested food, *(ii)* the regenerative cells which renew the destroyed and dead epithelial cells through secretion or in the process of degeneration and *(iii)* the goblet cells which are of uncertain functions. Thus, there are following five major function of enteric epithelium:

(i) *to make digestive enzymes*

(ii) *to absorb the digested food*

(iii) *to produce new cells*

(iv) *to absorb the water*

(v) *to excrete the waste material out side the body*

The inner surface of midgut is sometime lined by a thin membrane known as peritrophic membrane, the main function of which is to protect the epithelial cells from the direct contact of food particles. This membrane is absent in case of lepidopterans and hemipterans.

Hind Gut or Proctodaeum: It extends from the posterior end of midgut to the anus and is also an invagination of the body wall. The hind gut consists of the same layers as the forc gut except that the circular muscles of its are developed both inside and outside the layer of longitudinal muscles. The hind gut is externally marked by the insertion of the Malpighian tubes and internally by the pyloric valve. It may be divided into three distinct regions:

(i) *Ileum or small intestine*

(ii) *Colon or large intestine*

(iii) *Rectum*

Ileum: It is a small tube which has many folds in its inner wall.

Colon: It is situated on the 5th and 6th segments of the abdomen and is a slender tube which can not be easily distinguished from the ileum. In some insects it is just like 'S' in structure.

Rectum: Both the ends of the rectum are comparatively slender while the middle portion is thick and large which consists of six rectal papillae internally and six ridges of longitudinal muscles externally. The rectum opens to exterior through the anus which is situated at the caudal end of the abdomen.

Salivary Glands: The labial glands which are associated with the gnathal appendages are the salivary glands. A pair of salivary glands is found in the grasshopper which generally lie in the thorax and are convoluted tubes often branched and racemose. Both the ducts of salivary glands unite together beneath the oesophagous to form a common salivary duct which opens into the salivarium.

Physiology of Digestion

The grasshopper eats the leaves and soft parts of the plants which are hold by the maxillae and they bring the food near to mandibles where it is broken into small particles. These small food particles are sent to the buccal cavity with the help of labrum and labium. On entering the buccal cavity, it is subjected to the action of saliva which contains the amaylase enzyme. It acts on the carbohydrates present in the food and change them into simple sugar, *i.e.*, glucose which is absorbed in the crop. Saliva is also helpful in moistening the food. This food passes onward to the crop where the secretions of the midgut and the hepatic caecae mix with it. These secretions are weakly acidic or alkaline and contain maltase, invertase, lactase, protease, lipase, peptidase, erypsin and trypsin enzymes which act on the food.

Due to the action of these enzymes the starch is converted into sugars, protein into amino acids and fat into fatty acids. After this the food comes to gizzard where it is again masticated then it passes through the cardiac valve into mesenteron where further digestion of the food takes place. The digested food is absorbed by the spongy and thick walls of mesenteron. The undigested food passes to the hind gut (proctodaeum) through pyloric valve where the absorption of water takes place and then waste and undigested food expelled out through anus in the form of excreta. The absorbed food is utilized for the following purposes:

(i) In the form of energy required for different life activities.
(ii) Some part is consumed in the formation of muscles etc.
(iii) The rest is stored in the fat bodies which is used in emergency.

Excretory System

The function of the excretory system is to maintain a healthy internal environment for the proper functioning of tissues and cells by the regulatory processes involved in the elimination of metabolic waste products such as urea, CO_2 excess of water and other organic and inorganic matters. The principal excretory organs of insects are the Malpighian tubes, but accessory excretory function is also performed by the fat-body and nephrocytes to some extent.

Malpighian Tubes

These were first discovered by the Italian scientist Malpighi and the name was given after him. These organs are almost universally present among insects which are long, slender, blind tubes lying in the haemocoel where they are freely bathed by the blood. They open at their proximal ends into the intestine, near the junction of midgut and hind gut. Their number vary according to the species of insects. Generally, they are found

24 in number and are considered as derivatives of the ectodermal proctodaeum but Hensen considers them outgrowths of the mesenteron. Each tube may be divided into two parts: *(i)* proximal end *(ii)* distal end. The function and structure of these parts are different from each other. The distal end acts as excretory, *i.e.,* it excretes the nitrogenous waste products produced during metabolism while proximal end absorbs the inorganic salts (potassium carbonate, sodium carbonate etc.) and water. Thus, in proximal end, the precipitation of uric acid takes place. Finally the absorbed waste products are left in the hind gut.

As blood flows over the tubules, the cells of these tubules absorb the uric acid from the blood and discharge it into the lumen of the tube in aqueous solution. From this point uric acid solution (urine) is forced into the ileum and expelled through the anus in the form of crystals (urate).

Fat Body

The fat body is composed of irregular masses or lobes of rounded or polyhedral cells which are usually vacuolated inclusions and contain inclusions of various kinds. The colour of fat body may be either white, yellow, orange or greenish. The cells of the fat body seem to be closely related to the haemocytes. Generally fat is deposited in these tissues but in some insects deposits of uric acid or urates appear in the fat body. These excretory granules are formed in specialized cells of fat body known as urate cells. The excess quantity of waste materials which are not excreted by Malpighian tubules are deposited in these cells.

Nephrocytes

The nephrocytes consist of scattered or localized groups of cells and generally they occur in two principal groups:

(i) the dorsal or pericardial nephrocytes *(ii)* the ventral nephrocytes. The dorsal nephrocytes consist of two chains of cells arranged in a linear series one on either side of the heart in the pericardial sinus. Such cells are present in most of the

insects. The ventral nephrocytes occur in Dipterous larvae and usually form a chain of cells which are suspended in the body cavity below the fore intestine and attached by its two ends to the salivary glands. The nephrocytes are able to absorb colloidal particles.

Circulatory System

Among insects the circulatory system is of open type while in vertebrates it is closed. The arteries and veins are absent in insects. The greater part of the circulation of the blood takes place in the cavity of the body known as haemocoel and its appendages. The blood of insects is called as haemolymph.

Blood

The blood of insects is contained in the general body cavity, where it freely bathes the various internal organs and also enters appendages and the tubular cavities of the wing veins. The insect blood is usually yellowish or greenish liquid and called as haemolymph. It is slightly acidic in nature but the pH varies with the instar, age and sex of insect. Its specific gravity is close to that of water. It contains following things:

(i) Liquid matter (ii) Solid matter and *(iii) Living cells*

(i) *Liquid Matter:* It is known as haemolymph which is non-living. It contains certain pigments which are blue, yellow or green in colour and responsible for the colour of blood. There is about 85% water in blood.

(ii) *Solid Matter:* It includes inorganic ions, amino acids, proteins, fats, sugars, organic acids and other substances in variable quantities. The compounds of calcium, magnesium and phosphorus are present in small amount. When the blood of insect comes into contact of air it becomes dark green in colour, it is due to the oxidation of albuminoides and precipitation of uranidine into green and black granules. According to Poulton and associates, the green yellow pigments are received from the food which are not changed even after absorption. As the pigments are received from the food. The other materials such as sugars organic acids, amino acids, salts of potassium are also present. All these materials are responsible for the manufacturing of haemocyanin, chlorophyll pigments etc. The following are the important albuminoides found in the blood:

(i) *Haemoprasin* (ii) *Haemochlorin* (iii) *Haemopyrhin* (iv) *Haemocrosin*

Living Cells: The blood cells are found freely suspended in the blood and are known as haemocytes which are of following types:

1. *Proleucocytes:* These are small, spindle shaped and have bigger nucleus and basiphilic cytoplasm. They undergo a mitotis cell-division. Some people are of the opinion that these cells are the basic structures of all the blood cells.
2. *Phagocytes:* These cells differ from proleucocytes in having a bigger size but containing small nucleus. They ingest generally all the dead cells. Injurious microorganisms and some living cells also collect at wounds and form a plug to close the breaks in the body.
3. *Granulocytes or Spheroidocytes:* These cells are bigger in size and contain granular cytoplasm. The exact function of these cells are not known. According to some scientists they suck the food and some act as phagocytes.
4. *Oenocytes:* These are oval or round in shape and having light yellow homogenous cytoplasm.

Function of the Blood

The following are the known functions of the insect blood:

(i) *Transportation:* The digested food is absorbed from the digestive system and conveyed to the tissues where as the waste products are carried from the tissues to the excretory system.

(ii) *Respiration:* The cells not having tracheoles certainly receive their oxygen from the blood, thus it helps in respiration.

(iii) *Protection:* The haemocytes (phagocytes) dispose of injurious bacteria, parasites and dead cells.

(iv) *Hydraulic Function:* As stated earlier the entire volume of blood is enclosed within the body wall which forms a hydraulic system capable of transmitting pressure from one part to another part of the body.

Blood Circulatory System

Among insects, the circulatory system is largely an open type and blood flows in the body cavity (haemocoel). The haemocoel contains two fibro-muscular membranes, one of which is situated beneath the heart known as dorsal diaphragm while another is found in between the alimentary canal and nervous system called as ventral diaphragm. Both these diaphragms divide haemocoel into three sinuses:

(i) *Pericardial or Dorsal Sinus:* This is an upper chamber in which heart and aorta are situated.

(ii) *Middle or Perivisceral Sinus:* This is the middle chamber and alimentary canal is situated.

(iii) *Ventral or Perineural Sinus:* This chamber is situated beneath the ventral diaphragm in which nervous system is found.

Dorsal and ventral sinuses are not completely separated diaphragm but in many places small holes are found by which blood from middle sinus passes to ventral sinus. The following organs are involved in blood circulation:

(a) Dorsal vessel

(b) Dorsal and ventral diaphragm

(c) Accessory pulsatory organs

Dorsal Vessel: It extends from near the caudal end of the body, through the thorax into the head and lies beneath the dorsal body wall. It may be divided into two parts:

(i) The posterior abdominal portion called as heart.

(ii) The anterior thoracic portion the aorta.

(i) *Heart:* It runs as a slender tube, some what swollen in each segment, closely attached to the ventral side of the ceiling of haemocoelomic cavity. The posterior end of this tube is closed. The wall of the heart is composed of two layers (i) outer layer which is known as peritoneal layer or connective tissue and (ii) the inner muscular layer. A series of triangular muscles are found on each side of the heart which are known as alary muscles. These muscles extend from the heart to the lateral walls of the body and form the dorsal diaphragm.

Every swollen portion (chamber) of the heart is called as the ventricle, the posterior portion of which is divided into three lobes, one dorsal and two lateral lobes. A slit like opening is found at the base of each lateral lobe which is known as ostium through which the blood from hoemocoel enters the heart. Each ostium is provided with an ostial valve which permits the blood into heart but does not allow it to go back into the haemocoel. An auricular valve is found at both the ends of each ventricle.

(ii) *Aorta:* It is simple tube which extends from the Ist abdominal segment to head where it opens near the brain. Its lateral and dorsal sides are attached to the hypodermis. It functions as the principal artery of the body.

(b) *Dorsal and Ventral Diaphragm:* The dorsal diaphragm is made up of hard muscles having many small holes through which blood passes. Ventral diaphragm is not found in all insects. In grasshopper some muscle bands are developed over the ventral nerve cord which form the ventral diaphragm. It produces a flow of blood posteriorly and laterally by expansion and contraction.

(c) *Accessory Pulsating Organs:* They are sac-like structures situated in various regions of the body and pulsate independently of the heart, ensuring an adequate circulation of blood through the appendages. In grasshopper the cephalic ampullae are found in the anterior part of the head, near the base of antennae which drive blood into them. Among Hemiptera special pulsatile organs are present in the legs.

Physiology of Circulation: The heart is the principal pulsatory organ. The pulsation of the heart is caused by the alary muscles by their expansion and contraction. The two activities systole and diastole are also regularly performed in the heart of insect.

Systole: Due to the contraction of alary muscles, the paired ostial valves at the anterior end and the auricular valve at the posterior end of the ventricle are closed, therefore, the blood is to go forward into the next ventricle through the anterior auricular valve which is opened.

Diastole: Diastole results from the relaxation of the heart muscles and is immediately followed after systole. During this process the anterior auricular valve is closed while the ostial valves and the posterior auricular valve remain open, the ventricle which will undergo diastole will receive the blood supply from the pericardial sinus through ostial valves and from the adjoining hind ventricle undergoing systole in that duration through auricular valve.

Each chamber contracts one after another starting from the caudal end of the heart forwards thus a regular circulation of blood is maintained. From the anterior most ventricle, the blood undergoes to dorsal aorta and due to the closing of valves it is prevented from returning, goes to head.

Respiratory System

The lungs are absent in insects, therefore, the respiration takes place by means of internal air tubes known as trachea floating in the blood of haemocoel. The fine branches of trachea are termed as tracheoles. The respiratory organs of insects are always derived from the ectoderm. The tracheae are developed from solid ingrowths or tubular invaginations of the body wall. Histologically, therefore, composed of a layer of cuticle, the hypodermis and usually a basement membrane. Generally, most of the cuticular lining of the trachea is shed at ecdysis.

The air generally enters the tracheae through paired and usually lateral openings known as spiracles or stigmata. They are segmentally arranged along the thorax and abdomen and vary in size, shape and structure. A typical annular surrounds its opening known as permitreme. Besides, there is atrium or vestibule into which the opening leads. The atrium is a specialized region leading from the spiracular opening, it lacks taenidia and its walls are generally sculptured or provided with hairs. These help to reduce water loss and prevent the entry of dust.

In grasshopper, ten pairs of spiracles are present out of which two pairs are placed on the lateral sides of thorax and eight on the abdomen. According to the number and arrangement of functional spiracles it is possible to classify respiratory systems as follows:

(i) *Holopneustic (Holo = all or whole, pneustic = to breathe) :* In this case 10 pairs (2 thoracic and 8 abdominal) of functional and open spiracles are present, *e.g.*, grasshopper.

(ii) *Propneustic (Pro = first or before):* In which the anterior most pair of spiracles is functional and rest are closed *e.g.*, in case of mosquito pupa.

(iii) *Metapneustic (Meta = after)*: Only the last pair of abdominal spiracles is open, *e.g.*, in the larva of a mosquito.

(iv) *Amphipneustic (Amphi = two or both):* In which the first and last pairs of spiracles are functional, *e.g.*, in maggot of housefly.

(v) *Peripaeustic* (peri-around or about): In which all the spiracles except any one or two pairs are functional, *e.g.*, in grubs.

The respiratory system of grasshopper may be divided into following heads:

1. Longitudinal tracheal trunks.
2. Transverse tracheal trunks.
3. Air sacs.
4. Tracheoles.

Longitudinal Tracheal Trunks: There are three pairs of longitudinal trunks present namely:

(a) *Lateral tracheal trunks*

(b) *Dorsal tracheal trunks*

(c) *Ventral tracheal trunks*

Lateral Tracheal Trunks: These trunks are situated on the lateral sides of the body in which all the spiracles of the same side arc opening thus the spiracles of one side join together to the other side. The lateral tracheal trunks open in a large thoracic sac in the thorax and posteriorly it divides into two branches. The first branch is called as dorsal trachea which joins the dorsal tracheal trunk and the other is ventral trachea which joins the ventral trunk.

Dorsal Tracheal Trunks: The dorsal segmental trachea join on either side to form a pair of dorsal tracheal trunk. The trunks are placed parallel on both sides of the heart extending from Ist abdominal segment to eighth segment. The anterior and posterior ends of each trunk join with the lateral trunk of its side. Both the dorsal tracheal trunks are connected with each other at each segment by a transverse tracheal tube called as dorsal commissure.

Ventral Tracheal Trunks: The ventral tracheal trunks lie on the ventral side of the body and run parallel to the ventral nerve cord. They are also connected with the lateral trunks of their corresponding sides by transverse tubes in each segment of the abdomen. Both the ventral trunks are segmentally connected with a transverse ventral commissure and extend from the head to the eighth abdominal segment.

Transverse Tracheal Trunks: These are also called as branches of lateral tracheal trunks and go to the following parts:

(a) To alimentary canal — one branch in each segment goes to the alimentary canal. This branch divides into two and spreads on the upper and lower sides of the alimentary canal.

(b) To air sacs — One branch goes to each air sac.

(c) To wings and legs —Many branches lead to the wings and legs in the thorax.

(d) One branch goes to the dorsal side and joins the dorsal tracheal trunk in each of the first to eighth abdominal segments.

(e) Similarly, one branch goes to the ventral side and joins the ventral tracheal trunk.

Air Sacs: The air sacs are the dilated or inflated parts of the trachea and have close connection with the tracheal system. There is a pair of large sacs behind the eyes and several smaller ones in the region of mouth parts. In thorax, 4 pairs of sacs are found while in the abdomen eight pairs of large sacs are present. The sacs of each side are connected by a longitudinal tube running along the alimentary canal. Besides, there are numerous smaller sacs and inflated section of tracheae are present. The principal function of the air sacs is a respiratory one as they serve to increase the volume of 'tidal air' which is changed when respiratory movements are made.

Tracheoles: The tracheoles are slender, unicellular in origin and connected with the tip of trachea arising from their sides. They may contain liquid or air, and end blindly or in anastomosis with each other and their walls are permeable to water.

Physiology of Respiration

The entry of oxygen into the tracheal system and diffusion of carbon dioxide outside the body is known as respiration. To take the oxygen in the body is called inspiration while to diffuse carbon dioxide out side the body is expiration. During respiration all the spiracles do not function in the same way. In grasshopper during inspiration, 4 pairs of anterior (2 thoracic + 2 abdominal) spiracles are opened while remaining six are used during expiration. Since the walls of tracheae are flexible, these are controlled by blood pressure. Thus when the abdomen contracts the first four pairs of spiracles close automatically and the air is expelled through the remaining abdominal spiracles. Similarly, when abdomen expands, all the latter spiracles (6 posterior abdominal) close and inspiration occurs through the first four pairs. The air enters from them into the spiracular tracheae and then is driven through them into the other tubes.

According to Wigglesworth the gaseous respiratory exchange between the tracheoles and the tissues depends on diffusion. The tracheoles may be entirely filled with gas or contain liquid in their terminal parts; the amount of liquid in the tracheole is affected by the osmotic pressure of the surrounding fluid. Thus, in active muscle with an adequate oxygen supply the accumulation of metabolites raises the osmatic pressure of the tissue fluids; liquid is withdrawn from the tracheoles through colloid imbition and is replaced by air so improving the supply of oxygen to these tissues.

According to Imms the diffusion process which accounts for the inward transport of oxygen also permit the outward movement of carbon dioxide but the tissues and cuticle are far more permeable to carbon dioxide than to oxygen so that an appreciable proportion of the former can escape through the tracheal walls and the general surface of the body. The blood plays relatively minor role in respiration.

Kelin (1925) stated that the cytochrome, a intercellular respiratory pigment, is found in insects which is an important constituent of respiration.

Nervous System

In insects the nervous system is highly developed and serves as connecting link between the sense organs and the effect of organs (muscles, glands etc.). The nervous system of insects may be divided into three parts:

1. Central nervous system
2. Sympathetic or Visceral nervous system
3. Peripheral nervous system

Central Nervous System: This constitutes the main division of the nervous system and is composed of a double series of ganglia which are joined together by means of longitudinal and transverse strands of nerve fibres. The longitudinal cords are termed connectives and they serve to join a pair of ganglia with these which precede and succeed it. The transverse fibres or commissures unite the two ganglia of a pair.

The central nervous system is divisible into the following:

(a) *The brain or cerebral ganglion*

(b) *The suboesophageal ganglion*

(c) *The ventral nerve cord*

Brain: It is situated just above the oesophagous between the supporting apodemes of the tentorium. It is formed by the fusion of first three neuromeres (optic, antennary and inter calary ganglia) in the embryo. This three fold division is maintained which are designated as follows:

(i) *Fore brain or Protocerebrum*

(ii) *Mid brain or Deutocerebrum*

(iii) *Hind brain or Tritocerebrum*

The protocerebrum represents the fused pair of optic ganglia. It forms the greater part of the brain and innervates the compound eyes and ocelli.

The deutocerebrum is formed by the fusion of antennary ganglia of the antennary segment which innervates the antennae.

The tritocerebrum is formed by the ganglia of the third of intercalary segment of the head. It is divided into two small widely separated lobes which are attached to the dorsal lobe of the deutocerebrum and receive nerve fibres from the latter. The tritocerebrum lobes are joined together by means of postoesophageal commissure which passes immediately behind the oesophagous. They also give origin to:

(i) The para oesophageal connectives or crura cerebri which unite the brain with the sub-oesophageal ganglion.

(ii) The labro-frontal nerves which go to labrum and frontal ganglion.

Sub-oesophageal Ganglion: It is the ventral ganglionic centre of the head and is formed by the fusion of the ganglia of the mandibular, maxillary and labial segments. It joins the brain by a pair of large connectives known as sub-oesophageal connectives. These connectives pass on each side of the oesophagous and are termed as crura cerebri of the legs or the brain. The sub-oesophageal ganglion gives rise to the nerve trunks servicing the mouth parts. It is joined to ventral nerve cord by the connective passing through the neck.

Ventral Nerve Cord: It consists of a series of ganglia lying on the floor of the thorax and abdomen. They are united into a longitudinal chain by means of a pair of connectives which issue from the posterior border of the sub-oesophageal ganglion. The first three ganglia are situated on three thoracic segments and are known as thoracic ganglia, rest of ganglia are located in the abdomen and form the ganglia of that region. The thoracic ganglia control the locomotory organs. Each ganglion gives off two pairs of principal nerves, one of which supplies the general musculature of the segment and the other innervates the muscles of the legs. Similarly, there are five paired ganglia in the abdomen but the last ganglionic mass of the abdominal region is formed by the fusion of last three primitive ganglia and chiefly innervates the reproductive organs.

Each abdominal ganglion gives off a pair of principal nerves to the muscles of its segment.

The Visceral Nervous System: In the anterior portion of the alimentary canal and dorsal blood vessel, insect possesses a nervous system called as visceral or sympathetic nervous system. The central structure of this nervous system constitutes the frontal ganglion located in front of the brain and connected to the tritocerebrum by a pair of fibres. Anteriorly, it gives off a frontal nerve which passes to the clypeus.

Posteriorly, the frontal ganglion gives off a recurrent nerve which extends along the mid-dorsal line of the oesophagous and, passing just beneath the brain, expands a short distance behind the latter centre into a hypocerebral ganglion. This group innervates the stomodaeum, salivary ducts, aorta and apparently certain muscles of the mouth parts.

Peripheral Nervous System: This includes all the nerves radiating from the ganglia of the central and sympathetic systems. Just beneath the hypodermal layer of the body wall, there are many bipolar and multipolar nerve-cells whose prolongations form a net work of nerves. These nerves are named on the basis of a particular organ where they go, *e.g.*, eyes and ocilli nerves, mandibular, maxillary and labial nerves, leg nerves and wing nerves etc.

Various Organs

REPRODUCTIVE ORGANS

Basically, the insects are dioecious, *i.e.*, only one sex is represented in one individual. In their embryonic condition they are initially similar in sexes and later differentiate in their development. The reproductive system of both the sexes in general has been described below:

MALE REPRODUCTIVE ORGANS

The male reproductive organs consist of the following:

(i) A pair of testes (ii) A pair of vasa deferentia

(iii) Seminal vesicles (iv) Ejaculatory duct

(v) Penis or Aedeagus (vi) Accessory glands

(vii) Male genital atrium

The Testes: They are located above the midgut and held in position by the surrounding fat bodies and tracheae. Each testis is a more or less ovoid body partly or completely divided into a variable number of follicles or lobes which are cylindrical in shape. Each follicle is connected with vas deferens by a relatively well developed slender tube known as vas efferens. The peritoneal investment of the follicle is developed to the extent of enveloping the testis as a whole in a common coat known as scrotum.

The testicular follicles are lined with a layer of epithelium whose cells rest externally upon the basement membrane and outside of which there is a peritoneal coat of connective tissue. Each follicle may be divided in to a series of zones characterized by the presence of the sex cells in different stages of development, These zones are as follows:

(i) *The Germarium:* It is the region having primordial germ cells or spermatogonia which undergo multiplication.

(ii) *The Zone of Growth:* In this zone the spermatogonia increase in size and undergo repeated mitotic division and develop into spermatocytes.

(iii) *The Zone of Division and Reduction:* Here the spermatocytes undergo meiosis and produce spermatids.

(iv) *The Zone of Transformation:* The spermatids are transformed into spermatozoa.

The masses of spermatozoa are generally enclosed in the testicular cyst cells from which they are released in the vas deferens. In addition, the testes contain large elements known verson's cells or apical cells.

Vas Deferens: These are the paired canals leading from the testes which are partly or wholly mesodermal in origin.

Seminal Vesicles: The vasa deferens vary greatly in length in the majority of insects. Each vas deferens becomes enlarged along its course to form a sac known as seminal vesicle in which spermatic fluid is collected.

Ejaculatory Duct: Posteriorly, the vasa deferentia unite to form a short common canal which is continuous with a median ectodermal tube known as ejaculatory duct. The terminal end of ejaculatory duct opens in the male genital atrium.

Aedeagus: The terminal end of the ejaculatory duct is enclosed in a finger-like evagination of the ventral body wall which forms the male intromittent organ known as aedeagus. It is situated on 9th abdominal sternum of the grasshopper on the conjunctival membrane of the posterior margin.

Accessory Glands: These are one to three pairs in number and usually present in relation with the genital ducts opening into seminal vesicle. These are tubular or sac-like in structure. Very little information is available about their functions. In most of the cases their secretions mix with spermatozoa and in some insects glands are directly concerned with the formation of the spermatophores.

The Female Reproductive Organs: The female reproductive system consists of the following organs:

(i) A pair of ovaries
(ii) A pair of lateral oviducts
(iii) Spermatheca
(iv) Vagina and genital chamber
(v) Accessory glands (Collatrial glands)

The Ovaries: These are typically more or less compact bodies lying in the body cavity of the abdomen on either side of the alimentary canal. Each ovary is about 2 cm long and composed of a variable number of ovarioles and open into the oviduct. A typical ovariole is an elongated tube in which the developing eggs are disposed one after the other in a single chain. The oldest oocyte is situated nearer the union with the oviduct. The wall of an ovariole is made of follicular epithelium whose cells rest upon a basement membrane known as tunica propria.

Each ovariole may be differentiated into three zones:

(i) *Terminal Filament:* It is the slender thread like apical prolongation of the peritoneal layer. The filaments of the ovary combine to form a common thread termed as terminal filament. The terminal filament of one ovary units with the filament of the other ovary to form a median ligament. It aids in maintaining the ovaries in the position and is attached to the dorsal diaphragm.

(ii) *The Germarium:* It is situated below the terminal filament and forms the apex of an ovariole. It consists of a mass of cells which are differentiated from the primordial germ cells.

(iii) *The Region of Growth:* It is also called as vitellarium which constitutes the major portion of an ovariole. The vitellarium contains the developing eggs (oocytes). The epithelial layer of the wall of vitellarium grows inwards to enclose each oocyte in a definite sac known as follicle. The cells of the follicle secrete the chorion of the egg and in some cases serve to nourish the oocytes.

Three types of ovarioles may be recognised on the basis of presence or absence of nutritive cells.

(a) *Panoistic Type:* Nutritive cells are absent, *e.g.*, grasshopper and other insects of Orthoptera and Isoptera.

(b) *Polytrophic Type:* Nutritive cells are present and arranged in alternate with the oocytes, *e.g.*, Hymenoptera.

(c) *Acrotrophic Type:* Nutritive cells are present and situated at the apices of the ovarioles, *e.g.*, Hemiptera.

The Oviducts: The lateral oviducts are paired canals leading from the ovaries and are formed from the mesoderm. These lateral oviducts form the common oviduct which opens into the vagina. Each oviduct is an enlarged pouch which stores eggs. The vagina is greatly enlarged to form a chamber, known as uterus, for the reception of developing eggs.

The Spermatheca: This is a pouch or sac for the reception and storage of the spermatozoa (seminal fluid) and is also known as receptaculum seminis. It generally opens by a duct into the dorsal wall of the vagina which is known as sperm duct.

In many insects pairing takes place only once and since the maturation of eggs may extend after the union of the sexes, the provision of spermatheca allows for their fertilization from time to time. A special spermathecal gland opens into the duct of spermatheca and secretes a fluid which lengthens the life of sperms.

Genital Chamber: The vagina opens into the genital chamber on IXth sternum and this chamber is called bursa copulatrix which helps in copulation.

Accessory Glands: These are paired structures opening into the distal portion of the vagina. These glands provide material for the formation of egg pod or ootheca.

Fertilization: After copulation, the spermatic fluid is received in the spermatheca. The egg comes down from the oviduct to the vagina which has an opening (micropyle) into its shell for the entrance of male germ cell (spermatozoan). One or two spermatozoa enter the egg through micropyle and only one succeeds in fertilizing the egg. After fertilization the accessory glands secrete a fluid around the egg which hardens it.

Types of Reproduction: Generally the insect eggs are fertilized within the body of the female and deposited outside for further development. Besides, generalised rule there are certain specialized cases which are as follows:

Viviparous Reproduction: In this case, the insects instead of laying eggs, produce either larvae or nymphs, thus embryonic development is completed within the body of the female. This is a common parasitic Hymenoptera. It can be further categorised into four as below:

(a) *Ovoviviparity:* This is a condition where the eggs are retained in the genital tracts so that when they are laid, the embryos have reached the advance stage of development with the result the larvae enough yolk to nourish the developing embryos. This type is found in the representatives of the Thysanoptera, Blattidae, Muscidae, Cleoptera etc.

(b) *Adenotrophic Viviparity:* In this case the eggs are retained in the enlarged vagina (uterus) and the embryo develops inside egg and larva on hatching is retained in the uterus where it is nourished by secretion of accessory glands. The larva when fully developed is laid outside for pupation. It is found only in the diptera pupipara and in Glossina.

(c) *Haemocoelous Viviparity:* In this case there are no oviducts and the ovaries lie free among the fat body, breaking up readily when mature so that the eggs are deposited in the haemocoel. The eggs have no chorion but become surrounded from an early stage by a trophic membrane through which nutrient materials are supplied from the maternal tissue. The development of embryo occurs in the haemocoel of the mother insect. This type is found through out the order strepsïptera and larval cecidomyids.

(d) *Pseudoplacental Viviparity:* Here the embryo develops in an enlarged part of the maternal vagina from a practically yolkless egg which is almost always devoid of a chorion and, when no parthenogenetic, is fertilized in the ovariole. The embryo is nourished through placenta — like structures which are in close contact with developing embryo. The larva does not feed orally but derive nourishment through these pseudo-placentae. This type of reproduction is found in Aphididae, *Hemimerus* (Dermaplera), *Diploptera* (Blattidae), *Hesperoctenes* (Heteroptera) etc.

Parthenogenesis: Parthenogenesis is the development of the egg without fertilization. In some species it is of common occurrence while in others it is sporadic. It has not been observed in the odanata and Heteroptera. It may be of following types:

(a) *Haploid Facultative Arrhenotoky:* This type of parthenogenesis constitutes a sex-determining mechanism, which may be more flexible than the usual chromosomal type. Females lay fertilized (diploid) eggs which give rise to females and unfertilized (haploid) eggs which develop parthenogenetically into males, e.g., Hymenoptera, some Aleyrodids and Thysanoptera.

(b) *Facultative Thelytoky:* In this case oogenesis is accompanied by meiosis and if the eggs are fertilized they produce both males and females. The nuclei of those eggs which are not fertilized return to the deploid condition by fusion with the second polar body and develop only into females. It is well seen in Coccus hesperidum.

(c) *Obligate Thelytoky:* In this case males are generally absent or extremely rare, atleast sometimes nonfunctional. The eggs are often formed without meiosis or, when this occurs, there is a later doubling of the chromosome number, *e.g.*, by fusion of cleavage nuclei and some species with this type of parthenogenesis have achieved polyploidy in the germ line. It occurs in some curculionidae, many psychids.

(d) *Cyclical Parthenogenesis:* In some insects like aphids cecidomyids, Hymenoptera and others asexual reproduction cycle is interrupted by the sexual. It is often termed as alternation of generation where the insect derives the advantages of both parthenogenesis and sexual reproduction. In other words cyclical parthenogenesis can combine the genetic advantages of bisexual reproduction with the greater reproductive rate of thelytoky.

Paedogenesis: In some cases, immature insects (larvae and papae) possess functional ovaries and are capable of producing either the eggs or larval forms parthenogenetically. This is made of reproduction by the immature forms before attaining the adult stage is called paedogenesis. This situation is most common in *Micromalthus debilis* (Coleoptera) and in some cecidomyids.

Polyembryony: The term denotes the production of two or more (even very many) embryos from a single egg. In other words, it is the phenomenon where more than one embryo is produced from a single egg which divides repeatedly during the course of its development. It is quite common in a number of parasitic Hymenoptera belonging to the families chalcididae, Braconidae, Ichneumonidae etc.

Hermaphroditism: Functional hermaphroditism is (male and female organ in insect) an extremely rare phenomenon in insects, however, it is reported in the scale insect, *Icerya purchasi* where there are male forms which are extremely similar to the females of related species. In this case the gonad is formed by the anterior fusion of two originally separate organs, produces both spermatozoa and eggs. The outer cells of the gonad form ovarioles while the inner ones give rise to sperms.

Castration or Inhibition: Castration, in the broad sense, implies any process which inhibits completely or to a considerable extent the production of mature gametes by the organism, whose gonads may be greatly atrophied or well developed but not normally functional. In short it may be defined as "the conditions preventing maturation of gonads." It may be of two types:

(a) *Physiological Castration:* This is mainly due to nutritional deficiencies due to which proper development of gonads is inhibited as in the case of honeybee where the larval workers is dificient (royal jelly) in certain nutrients with the result that the adult worker though female, is non-functional because of undeveloped gonads.

(b) *Parasitic Castration:* The development of various parasites within an adult insect may induce sterility which is, in some cases, accompanied by changes in the secondary sexual characteristics of the host. This is most common in Aculeate Hymenoptera.

Sense Organs

The sense organs are those receptors whereby the energy of a stimulus arising outside or within the insect, is transformed into a nervous impulse. The transmission of nerve impulse to one of the central ganglia, usually results in a change in the behaviour of the insect or in the maintenance of some existing activity. In the broader sense, the sense organs or receptors may be divided into two classes.

(i) *Exteroceptors:* They perceive the stimuli arising in the external environment.

(ii) *Interoceptors:* They are excited by stimuli arising within the body.

On the basis of functions performed by the sense organs, they can be classified into the following main groups:

1. Photo-receptor organs
2 Chordotonal organs
3. Olfactory organs
4. Gustatory organs
5. Temperature and humidity receptors
6. Tactile organs

Photoreceptor or Visual Organs: The following are the photo receptors which are found in insects:

1. Dermal light sense
2. Ocellus or simple eyes
3. Compound eyes

Dermal Light Sense: It has been found that some insects react to light even after the removal of ocelli and compound eyes. In such cases, body surface seems to be sensitive to light but the localized receptors have not been identified so far.

Ocellus: Two types of ocelli are found in insects.

(a) Dorsal ocelli

(b) Lateral ocelli

Dorsal Ocelli: Typically, they are three in number and situated in a triangle. Generally, they are borne on the frons but in grasshopper, the median ocellus is located on the frons while the paired ocelli on the vertex. The following parts can be distinguished in an ocellus:

(i) *The Cornea:* It is a transparent cuticular structure which is arched or raised to form the outer investment of the ocellus.

(ii) *The Corneagen Layer:* This layer is continuous with the hypodermis but differs in being composed of colourless transparent cells. It secretes and supports the lens.

(iii) *The Retina:* It is composed of visual cells which are sensory neurones. The visual cells are associated with a group of cells and each group is termed as retinula which surrounds a longitudinal-optic rod known as rhabdom. The rhabdom is produced along the inner junctions of the component cells of a retinula.

(iv) *Pigment Cells:* In some cases the ocelli contain accessory cells with pigment situated between the retinulae.

The functions of the dorsal ocelli are not exactly known. It has been reported that blackening of the dorsal ocelli reduces the speed with which some insects respond to stimulation of the compound eye by light, therefore, lateral dorsal ocelli may be regarded as stimulatory organs.

Lateral Ocelli: They are also known as stemmata and are generally present in insect larvae. As their name indicates, they are situated on the side of head where they occupy positions corresponding with those of the compound eyes of the imagines. They differ from the dorsal ocelli in the fact that they are innervated from the optic lobes of the brain.

Compound Eyes: In compound eyes the cornea may be differentiated into a number of separate facets, thus differs from an ocellus which has only a single facet. Compound eyes are formed of aggregations of separate visual elements called as ommatidia. Each ommatidium representing with a single facet of the cornea and rests upon the basement membrane. The various parts which enter into the composition of an ommatidium are as follows:

(i) *The Cornea:* It is the transparent area of cuticle forming the facet or lens of an ommatidium and generally biconvex in form. It is shed during each moult.

(ii) *The Corneagen Layer:* It is the part of hypodermis which travels beneath the cornea and is known as the corneagen layer. It is composed of two cells which can not easily be detected. In cases where these cells are absent, the outer ends of the cells of crystalline cone secrete the cornea.

(iii) *The Crystalline Cone Cells:* In the Eucone eyes there is a group of four cells beneath the cornea or corneagen layer, which secrete a transparent body known as the crystalline cone.

(iv) *The Ratinula:* This structure forms the basal position of an ommatidium and is composed of a group usually seven pigmented visual cells. These visual cells secrete an internal optic rod or rhabdom. The portion of the rhabdom formed by each cell is termed as rhabdomere. The rhabdom forms the central axis of the ratinula and joins with the extremity of the crystalline cone.

(v) *The Primary Iris Cells:* These are deeply pigmented cells which surround the cells of the crystalline cone and the corneagen layer.

Rhabdomere forms the nerve fibre and the cells of ratinula rest on the basement membrane. From the posterior end of each cell a nerve fibre joins the optic nerve passing through the basement membrane. Thus a optic nerve is formed by the union of a number of nerve fibres. The image of the objects is invariably formed on the passing through the basement membrane. Thus a optic nerve is formed by the union of a number of nerve fibres. The image of the objects is invariably formed on the ratinula which is transferred to the brain by optic nerve.

(vi) *The Secondary Iris Cells:* These are commonly elongated pigment cells which surround the primary iris cells and the ratinula. Thus they serve to isolate an ommatidium from its neighbour so that there may not be any complication in image formation of an ommatidium.

Physiology of Vision: Each compound eye of an insect possesses thousands of ommatidia and each ommatidium acts as an eye then question arises whether an insect perceive thousand images of an object or only one. The answer of this question may be given with the mosaic theory of vision proposed by Muller, 1826.

According to this theory the rays passing through the cornea and crystalline cones reach the percipient portion of the eye. From each ommatidium the cornea transmits light to the crystalline cone from a very narrow field of vision and when this light comes to the crystalline cone, it forms an invariably a point of light and never an image. Therefore, the image formed upon the combined retinulae is a mosaic points of light and image thus formed is an erect one. Basically, there are two methods of image formation as described below:

(a) *Apposition Image:* This type of image is formed in the diurnal insects which possess a particular type of eye known as day eyes. In these eyes there is an envelope of pigments surrounding the transparent parts of each ommatidium in such a fashion so that only the light that has passed through cornea and crystalline cone (collectively known a dioptric unit) of that ommatidium reaches its rhabdom. Since the oblique rays of light are absorbed by the pigment cells hence, only the direct rays passing through the ommatidium forms the image on the distal portion of ratinula. Thus, the image is formed only of that portion of object which directly faces the ommatidium not in others.

In such cases the image is formed collectively by the different pieces of image of different ommatidia which are very close to each other. Therefore, this type of vision is also known as mosaic vision which forms a distinct image in the eyes of an insect.

(b) *Super Position Image:* This type of image is formed in the eyes of the nocturnal insects which are known as night eyes. Such eyes are meant to perceive objects and their movements in dim light. In this case the transparent part of each ommatidium is incompletely surrounded by an envelope of pigment and thus the rays of light traversing several neighbouring corneas can reach the same ratinula. Therefore, there will be an overlapping of the points of light and the image thus formed is known as super position image. It is evident that such an image is always a dim image not as distinct as an apposition one.

In many insects both types of vision, *i.e.*, apposition and super position types are found in a compound eye which is known as binocular vision. Such type of vision is found in mantis.

Chordotonal or Scolopophorous Organs: These are also known as auditory organs which are usually compound structures composed of a number of specialized sensilla (scolopophores). Each sensillum contains a conspicuous rod (scolopale or scolops). Chordotonal organs consist typically of a spindle-shaped bundle of scolopophores attached at each end to the integument or may end freely in body cavity. Each scolophore has a complex structure in which a bipolar neurone, produced basally into a fibre of the chordotonal nerve, is transformed distally into a slender prolongation covered by an envelope cell and cap cell. The scolopale is formed within an envelope cell and its cavity communicates basally with the liquid filled vacuole. These organs are generally found in the abdominal appendages and wing bases of insects. It is reported that besides sound receptors they are also used for various purposes including proprioception and the perception of internal pressure changes and mechanical vibrations.

The highly specialized type of chordotonal organs are Tympanal and Johnston's organs which are described as follows:

Tympanal Organs: Typically, these are paired structures made of a thin cuticular membrane known as the tympanum. They are associated with tracheal air-sacs and chordotonal sensilla. In grasshopper, tympanal organ is situated on each side of the first abdominal tergum and externally it is surrounded by a cuticular ring. Internally, there is a sensory organ known as Muller's organ which is formed by a group of numerous scolopophores and connected to the metathoracic ganglion by auditory nerve. Besides, there are two sclerotized processes and a pyriform vesicle filled with fluid. These structures are intimately associated with the Muller's organ and probably serve to transmit the vibration of the tympanum to the sensilla. These tympanal organs have reported to occur at the base of each fore tibia in the members of Tettigonoidea and Grylloidea.

Johnston's Organ: This is a highly specialized type of chordotonal organ situated in the second antennal segment of most of the insects. This organ was first seen and reported by Crystofer Johnston which consists of a variable number of radially arranged sensilla. These sensilla are attached at one end of the membranes between 2nd and 3rd antennal segments and at the other to the wall of the second segment. The exons from them enter the antennal nerve. The Johnston's organ stimulated by movements of the antennal flagellum which results in the movement of conjunctival plates located on either side of its base. This is how the sound is perceived by the sensilla attached with these plates. Thus, it is clear that this organ not only acts as a proprioceptor but also enables the insect to perceive air currents, vibrations of water surface, and contact with solid objects. It also acts as a sound receptor in culicidae family.

Olfactory Organs

The sense of smell is stimulated by low concentration of the vapour of substances which are relatively volatile at ordinary temperature. In honey bee, the olfactory pits are large in number which enable them to feel the sense of smell. Besides, olfactory hairs are also found in different parts, specially on the mouth parts and legs of the insects which help them in receiving the sense of smell.

Gustatory Organs

These are taste organs which respond to relatively high concentration of a stimulant in the liquid form. They are basiconic and short trichoid sensilla generally found on the antennae, surface of the preoral food cavity and mouth parts of most of the insects. In many members of Lepidoptera, Diptera and honey bees, gustatory organs have been reported to occur on the tarsus and distal parts of the tibia.

Their occurrence has also been seen on the ovipositor of some insects.

Humidity and Temperature Receptors: These are the most important organs specially meant for controlling the desiccation in insects. Humidity receptors have been identified in Pediculus in the form of a peculiar tuft like cuticular sensilla and in many species as basiconic, trichoid and placoid sensilla. With the help of these sensilla ants predict about the rains well in advance that is why they migrate to safer places alongwith their eggs, larvae and pupae prior to on set of monsoon.

Generally, temperature receptors, used for transforming heat to or from an insect are found on the antennae, maxillary palps or tarsi of the insects.

Tactile Organs

A large number of sensory hair are found throughout the insect body specially on the antennae, tarsi and cerci. They are stimulated to produce a nerve impulse on movement of hair in its socket. Each sensory hair comprises the trichogen and tormogen cells which secrete the hair and its socket, respectively together with a bipolar neurone.

External Morphology of Grasshopper: The grasshoppers are widely distributed through out the country and may be seen in abundance during monsoon season. For the generalised morphological description this insect has been considered as a most suitable representative of class Insecta because its structural details are not much more variable. Besides, being larger in size, it can be studied easily. The morphology of grasshopper has been dealt with as under:

Body Wall

The upper most layer of the body is known as body wall or the integument, which serves as an exoskeleton and the only alternative in insects in place of internal skeleton of vertebrates.

Typically the body wall of an insect is composed of following three distinct layers:

(i) *The cuticle*

(ii) *The epidermis or hypodermis*

(iii) *The basement membrane*

The Cuticle

It is an outer most layer of the body which is a complex, non-cellular layer mostly secreted by the hypodermis. It also serves a lining to the fore and hind intestine, tracheae and the other parts similarly formed by an ingrowth of the ectoderm. The cuticle consists of three layers:

(a) *Epicuticle:* It is an outer most layer measuring less than 4 m thickness and consists of hardened protein called cuticulin. Epicuticle is comprised of (1) Cement, (2) Wax, (3) Polyphenol and (4) Cuticular layers.

(b) *Exocuticle:* It is much thicker consisting mainly of chitin and proteins, the latter being 'tanned' by phenolic substances to produce a hard, brown material called sclerotin which gives rigidity to cuticle.

(c) *Endocuticle:* It is the inner most and thickest layer of cuticle which contains chitin and proteins. Since it is not 'tanned', therefore, becomes soft and flexible.

Both endocuticle and exocuticle consists of many laminae mostly arranged parallel to the surface and are acrossed by numerous pore canals containing thread like, cytoplasmic extension of the epidermis. The major component of the cuticle is chitin which makes up 25-60% of the dry weight of the cuticle. The chitin is a colourless nitrogenous polysaccharide, consisting of many sugar like residues and joined end to end in long molecular-chain.

It is insoluble in water, dilute acids, alkalies and organic solvents but dissolves in concentrated mineral acids. The protein component of the cuticle consists of two types of fractions in which one is water soluble and known as arthropodin and another has not been named and fully studied.

Functions of Cuticle

The main functions of cuticle are given below:

1. To give the proper shape to insect body.
2. To protect the internal organs of the body.
3. To check the water loss through the body surface.
4. To form the different organs inside the body, *e.g.*, tentorium, apodemes etc. During development period insect casts its cuticle and this process is known as moulting and casted cuticle as ecdysis. In place of old cuticle, new cuticle is formed by the hypodermis. The moulting is most essential phenomenon as in absence of this insect can not grow in size due to its rigidity. Therefore, as many times as insect has to grow, the cuticle is casted.

Processes of Cuticle: There are two types of processes formed by the cuticle:

1. The external processes include the setae, spurs, spines, bristle and hairs, which are the out growth of body wall.
2. The internal processes which are formed by the invasion of body wall, called apodemes. These structures provide internal areas for muscular attachment.

Epidermis or Hypodermis

It is the middle layer of the body wall and forms a single continuous layer of cells. The main functions of hypodermis are as follows:

1. It secretes the greater part of the cuticle.
2. It produces moulting fluid.
3. To absorb the digestive products of the old cuticle and to repair the wounds.
4. To form the epidermal glands.

The following glands are formed by the epidermis:

(i)	Salivary glands	(ii)	Moulting fluid glands
(iii)	Poisonous glands	(iv)	Scent glands
(v)	Wax glands	(vi)	Silk glands
(vii)	Mucous glands	(viii)	Lac glands
(ix)	Excretory glands	(x)	Repellent glands

Basement Membrane

It is the inner layer of body wall in which the inner ends of the hypodermal cells are attached. This is acellular thin membrane whose origin is uncertain but may be formed from blood cells.

The body of grasshopper is covered with hard plates which are known as sclerites. They are attached to each other with a thin membrane known as grooves and the joints are called sutures.

Body Regions

The insect body is composed of twenty primitive segments, which may be seen in embryo and grouped into 3 well defined regions ; the head, the thorax and the abdomen.

The Head

This is an anterior part of the body formed by the fusion of six segments viz., ocellary, antennal, intercalary, mandibular, maxillary and labial. All these segments are closely amalgamated to form a hard case or head capsule. The head generally assumes the following three definite positions with the body axis:

(i) *Hypognathus:* In this position the long axis of the head is vertical and mouth parts are ventral. The median line of the head forms right angle with the median line of the body, e.g., Grass hopper.

(ii) *Prognathus:* Here the long axis is horizontal and the mouth parts are anterior in position, so the median line of the head is parallel with the median line of the body, e.g., Beetles.

(iii) *Opisthognathus:* In this type, the head is directed back wards and mouth parts are posterioventral, e.g., Bugs.

Sutures and Areas of Head

The head capsule is demarcated by several sutures and the description of their details is as follows:

(i) *Frons or Front:* It is an area on the anterior face lies between or below the epicranial arms just below the vertex. It extends from the frontal suture to clypeus and up to the base of both the mandibles. The median ocellus is situated on sclerite.

(ii) *Clypeus:* It is a lip like area between the frontoclypeal suture and the labrum. The labrum or upper-lip is hinged with this by means of clypeo-lateral structure.

(iii) *Vertex:* The part of the head above the frontal suture is the vertex which lies above frons. The ocelli and antennae are situated in this region.

(iv) *Epicranial Suture:* It is an inverted Y shaped suture lies posterior between the eyes, whose stem begins on the back of the head, crosses the vertex and fork on the face. The stem of this suture is termed as epicranial stem or coronal suture and the two arms as epicranial arms or frontal suture.

(v) *Occiput:* This is situated in the back portion of the head and divided from the vertex and genae by occipital suture. The occiput lies between the vertex and neck.

(vi) *Genae:* These are the lateral walls of the head situated below and behind the eyes.

(vii) *Post Genae:* The areas directly posterior to the eyes are called post genae or areas of cranium posterior to the genae are termed as post genae. An occipital suture lies between the gena and post gena which completes the lower wall of the head with occiput in which Foramen magnum is situated.

(viii) *Compound Eyes:* These are paired structures situated in dorso-lateral position to head capsule. Each eye is surrounded by a narrow ring like ocular sclerite.

(ix) *Ocelli:* These are located in between the compound eyes and generally three in number, one of which is generally situated on the frons.

(x) *Antennae:* They are situated in antennal socket between the compound eyes. Each antennal socket is surrounded by a narrow ring like antennal sclerite.

(xi) *Tentorium:* The endoskeleton of head is known as the tentorium. It consists of paired anterior and posterior arms whose origins are visible externally as slit-like pits. The inner ends of the arms amalgamate to form the body of tentorium, from which a third dorsal pair of arms often arises. Thus, tentorium is composed of four principal parts namely, the anterior arms, posterior arms, corporatentorium or central mass and dorsal arms. The tentorium gives rigidity to the head capsule, provides attachment for muscles and supports the brain and oesophagous.

Appendages of Head

The head appendages include the antennae and mouth parts.

Antennae

They are paired freely mobile segmented appendages articulated with the head in front of or between the eyes.

Antennae are also known as feelers. They are multi segmented, but may be divided into three parts:

(i) *Scape:* It is basal segment of antenna, by which it is attached with the head.

(ii) *Pedicel:* It is the second segment of antenna which is shorter than scape. It bears sensory apparatus known as organ of Johnston.

(iii) *Flagellum:* The remaining divisions of antennae together constitute the flagellum which varies greatly in its form and structure according to the surroundings and habits of insects. This part is also known as clavola.

Functions of Antennae

The main function of antenna is sensory, which is modified according to use and need of insect as given below:

(i) *Organs of Smell:* In some insects the smell organs (sensoria) are situated in the antennae by which they recognise their food etc., e.g., ants, honeybee and giant moths.

(ii) *Organs of Taste:* Some insects bear taste hair on their antennae by which they know the taste of their food, e.g., cockroach.

(iii) *Stridulatorial Organs:* Sound producing organs are located in the antennae of some insects belonging to the orders — coleoptera and orthoptera, e.g., crickets.

(iv) *Chordotonal:* The hearing organs often known as Johnston's organs situated in the second segment (pedicel) of antenna, e.g., male mosquito, green bottle fly and paper wasp etc.

(v) *Sexual Characters:* In some of the insects belonging to the order Diptera and Hemiptera, the antennae are found of different type in male and female viz., Mosquito.

(vi) *Other Functions:* In some larvae the antennae are adapted for seizing the prey, e.g., chaoborus while in other insects they are used for holding the females, e.g., the male of Meloe. The butterflies are having some transmitting and receiving organs in their antennae.

Modifications of Antennae

On the basis of shape and structure, the antennae may be of following types:

(i) *Setaceous:* These are bristle like antennae in which the size of each segment becomes smaller and smaller towards flagellum and tappering into a point, e.g., insects of order Coleoptera, Lepidoptera and Orthoptera namely cockroach and cricket.

(ii) *Filiform:* It is thread like antenna in which all the segments are of nearly uniform in thickness and have no prominent constrictions at the joints, e.g., insects of order, Orthoptera. Coleoptera and Lepidoptera namely grass hopper and ground beetle.

(iii) *Moniliform:* All the segments of this antenna are globular in shape and of uniform thickness looking like a string of beads. There is a prominent constriction between the joints, e.g., insects of order Isoptera and Coleoptera namely termite.

(iv) *Serrate:* This is a saw like antenna in which each segment is more or less triangular, projected on one direction like the teeths of saw, e.g., Order Coleoptera (Pulse beetle).

(v) *Pectinate:* This is a comb like in structure in which each segment of the antenna possess long projection. Order Lepidoptera- moths (sugarcane root-borer).

(vi) *Bipectinate:* In this type of antenna each segment has the long projections on both the sides instead of one side like pectinate, e.g., Order-Lepidoptera (silk moth).

(vii) *Flabellate:* This is a comb like antenna in which the projections of some upper segments become long and form a feather (folding fan) like structure known as flabella. The flagellum seems to be lodged with this structure, e.g., order-Strepsiptera and Coleoptera-cedar beetle.

(viii) *Plumose:* In such type of antenna, the whorls of hairs arise from the joint of each segment and look like plume, e.g., order Diptera-male mosquito.

(ix) *Whorled:* Basically these are setaceous, filiform or moniliform types of antennae in which there is a whorl of bristle at every joint, e.g., order Hemiptera (Male of mango mealy bug).

(x) *Clavate:* These are club shaped antennae in which the segments become gradually broader towards the tip and the last segment finally ending into a round core, e.g., order Coleoptera and Lepidoptera-Butter flies.

(xi) *Capitate:* In this type of antenna, the terminal segment or segments form a large knob or cap, e.g., order Coleoptera and Lepidoptera — Necrobia.

(xii) *Lamellate:* These are the modifications of capitate antennae in which the terminal segments instead of forming the knob are extended on the side to form broad leaf like plate, e.g., order Coleoptera — Scarabaeid beetles.

(xiii) *Geniculate:* In this type of antenna, the first segment (scape) is long, second (pedicel) is short and flagellum is made of small segments which are bent on the scape just like a bent knee, e.g., order Hymenoptera — Honey bees.

(xiv) *Aristate:* This type of antenna is three segmented in which first segment (scape) is smaller and broader and the second (pedicel) is longer than the first. The flagellum is longer than both the segments and bears a heavy bristle known as arista e.g., order Diptera-House fly.

(xv) *Stylate:* In this antenna, the last segment of flagellum is modified into a long bristle known as style, e.g., order Diptera — Snipe fly.

(xvi) *Fusiform:* The basal and distal segments of this antenna are smaller and thin, while middle segments are larger just like a radish. The last segment is modified into a hook like structure. It appears like the keel of the boar from lower side, e.g., order Lepidoptera — Sphingid moth.

Mouth Parts

These are the organs primarily concerned with the uptake of food. They are the gnathal appendages and lobes of the head surrounding the oral aperture of the alimentary canal collectively known as mouth parts. A typical mouth parts, met in the insect, consists of the following parts:

(i) Labrum *(upper lip).*
(ii) A pair of mandibles.
(iii) A pair of maxillae.
(iv) Labium *(lower lip).*
(v) Hypopharynx *(tongue).*

The mouth parts of insects can be divided into different groups depending upon the type of food and method of feeding. The important modifications of insect's mouth parts are given below:

1. Biting and chewing type
2. Piercing and sucking type
3. Sponging type
4. Siphoning type
5. Chewing and lapping type
6. Rasping and sucking type
7. Degenerate type

Biting and Chewing Type Mouth Parts

These are typical type of mouth parts found in the insects belonging to the Order Orthoptera, Isoptera and Coleoptera. Besides, they are also found with slight modifications in the larvae of Lepidoptera, Coleoptera and Neuroptera. The insects having such type of mouth parts make holes on the leaves and cut it irregularly. The biting and chewing type of mouth parts consists of *(i)* labrum *(ii)* epipharynx *(iii)* a pair of mandibles *(iv)* a pair of maxillae *(v)* labium *(vi)* hypopharynx.

Labrum: The labrum is flap like structure attached to the clypeus and acts as an upper lip. It's important functions are to close the front of the mouth cavity, to protect the mandibles and to guide the food into the mouth. It has a notch at the distal end which helps in catching the leaves.

Epipharynx: It is not a well defined part in all insects, however, in some insects it is an important structure with the lower surface of labrum. The taste buds are located on this part.

Mandibles: A pair of mandibles is located directly behind the labrum. Each mandible is unsegmented, thick, strong, triangular in structure and its upper surface is flat. The mesal surface of a mandible is differentiated into a distal toothed incisor lobe and proximal molar lobe. They are hinged to the head by anterior and posterior articulations and move in lateral manner by abductor and adductor muscles. The mandibles are primarily meant for chewing and grinding the food.

Maxillae: These are also paired structure lying behind the mandibles. The maxillae are used for holding the food so that mandibles may perform their functions easily. Each maxilla has following well defined parts:

(i) *Cardo:* is a triangular basal part attached to the head capsule and serve as a base for the movement of maxilla.

(ii) *Stipes: is* rectangular, central part of the maxilla situated above the cardo. On the outer side of the stipes, there is a distinct sclerite known as palpifer to which the palpus is attached.

(iii) *Galea:* is the outer lobe, comparatively soft and has two segments. The basal segment is termed as subgalea or parastipes.

(iv) *Lacinia:* is the inner lobe of maxilla which is comparatively hard having a series of spines or teeth on its inner (mesal) edge.

(v) *Maxillary Palp:* Are antenna like appendages attached with the lateral side of the stipes. The portion of stipes to which the maxillary palp is attached known as palpifer. They are five segmented and sensory in function.

Labium: It is the lower lip of mouth and closes the mouth cavity from below or behind. Some times it is also termed as second maxilla. The labium appears to be a single organ, but rarely consists of two units which have fused to form a functional structure. It may be differentiated into the following parts:

(i) *Submentum*—is the flat leaf like basal part of the labium.

(ii) *Mentum*—is the central part of the labium, which bears a pair of three segmented palp on its either side. The portion on which palp is attached known as palpiger.

(iii) *Prementum* — is the upper portion of labium, which has four terminal lobes. The median pair is called as glossa and the lateral one as the paraglossa. These lobes are just like lacinia and galea of the maxilla and commonly known as ligula.

Hypopharynx: It is a tongue like structure arising from the floor of mouth cavity. The salivary glands open through it.

Piercing and Sucking Type Mouth Parts

These are highly specialized mouth parts which are primarily adapted for piercing and sucking the sap from plant or animal epidermis. They may be divided into the following subtypes of mouth parts depending upon their functions and mode of feeding:

(a) *Bug type.*

(b) *Mosquito type.*

Bug Type: Such type of mouth parts are found in Hemiptera which are used for sucking the sap from the plant system. These mouth parts consist of the following parts:

(i) *Labrum:* It is a small unsegmented triangular lobe attached with the clypeus from the base and anteriorly it covers the labial gutter. It is often called the epipharynx which is generally vestigeal in structure.

(ii) *Mandibles:* These are modified to form slender hollow bristle like stylets. In the inner surface of the mandibular stylet there is a long groove which helps in the upward and downward movement of the maxillary stylets. The anterior region of the mandibular stylets consist of recurved incisors which help in holding the plant tissues.

(iii) *Maxillae:* These are also paired needle like stylets which taper into a fine point and are grooved along their inner aspect. The groove of each maxillary stylet is divided into two parallel channels by means of a longitudinal ridge which runs from the mouth cavity to the end of the stylet. By the approximation of the maxillary stylet two extremely fine channels are formed, the dorsal one functions as the food channel and communicates with the sucking pump (cibarial), the ventral tube is the ejection canal or salivary channel which receive the saliva from the salivary duct. The maxillary stylets are devoid of the maxillary palp.

The movement of mandibular and maxillary stylets is governed by the alternate action of protractor and retractor muscles. During resting stage the stylets lie in the labial groove:

(iv) *Labium:* It is modified into a beak like structure named as rostrum which is three or four segmented, long, slender and rigid organ. On its dorsal side, there is a groove which acts as a resting site for the stylets. At the distal end of the rostrum, hairs are found. The labial palps are absent.

(v) *Hypopharynx:* It is highly specialized structure which is small and well sclerotized lying near the base of the stylets. Its median portion (Sitophore) forms a part of the floor or the cibarial sucking pump. Beneath it is the salivarium and salivary duct passes through it.

(vi) *Suction Pump:* Its upper wall is formed by the epipharynx and the lower wall by the hypopharynx. In fact it is the pre-oral cavity which is modified into the sucking pump and serves as a force pump and reservoir of saliva. Its working is governed by the dilator muscles which are attached with epipharynx.

Feeding Mechanism: This process may be explained in the following two phases:

(i) *Piercing Process:* At first the insect locates the suitable feeding site with the help of hairs situated at the distal end of the rostrum. After the proper selection of the site, mandibular stylets are penetrated into the plant tissues and thereafter the maxillary stylets are inserted into the plant material and with the looping of labium maxillary stylets get deeper and deeper in the plant tissues. During penetration of the stylets the saliva is injected through the salivary channel which mixes with the sap and forms a lubricated channel. This facilitates the movement of the maxillary stylets.

(ii) *Sucking Process:* When the piercing is over, the dilator muscles contract and upward suction is created through the food channel by capillary action as a result of that the sap is sucked up from the plant cells by the turgor pressure into sucking chamber. Thereafter the dilator muscles regain their original shape which creates the pressure in the sucking chamber and forces the food into the oesophagous.

Mosquito Type: Such type of mouth parts are primarily modified for blood sucking purpose and are commonly found in the female mosquito, fleas and certain blood sucking flies etc. The following are the distinguished parts:

(i) *Labrum-epipharynx:* The labrum and epipharynx are fused together to form a long and slender groove or 'U' shaped structure. In association with the flattened hypopharynx it constitutes a close tubular structure termed as food channel and through which the blood is drawn up by the pumping action of the cibarium.

(ii) *Mandibles:* These are paired structure modified into fine needle like stylets which are commonly known as mandibular stylets. They are comparatively weaker and smaller in male mosquito.

(iii) *Maxillae:* The maxillae also assume needle like structure and are termed as maxillary stylets. They are distally armed with minute teeth which are meant for making the wound in the skin of the host. These teeth are considered to be galea where as lacinia are absent. Unlike bugs, the maxillary stylets are provided with four segmented maxillary palp at their respective base.

(iv) *Hypopharynx:* This is a long, slender and double edged sword shape in structure. The salivary channel passes medially and conveys the saliva to the distal orifice of the 4 channel where it mixes with the blood. It is situated at the lower surface of the labrum-epipharynx and helps in forming the food channel.

(v) *Labium:* It is a well developed portion of the mouth parts and is modified into a thick elongated structure. Dorsally it is grooved and acts as a sheath to retain the hypopharynx and all the stylets when at rest. These appendages are finally covered by the labrum epipharynx. At the anterior end of the labium there are two lobes provided with many sensory hairs and known as labellae. At the proximal end of the labium there is a pair of three jointed maxillary palp.

(vi) *Sucking Pump:* The structure and function of the sucking pump of mosquito is similar as described in bug type of mouth parts.

Feeding Mechanism of Female Mosquito: With the help of abductor and adductor muscles the maxillary stylets penetrate the skin and then the mandibular and other stylets are also inserted into the puncture. During the feeding process the saliva is mixed with the blood which acts as anticoagulant, *i.e.*, it does not allow the blood to coagulate so that the insect may suck the blood easily. The piercing and sucking processes are the same as described for bugs.

Sponging Type Mouth Parts

This type of mouth parts are known as muscoid type and are commonly found in the Order Diptera (non-blood sucking flies). The classical example of such mouth parts is met in house fly. These are highly modified and primarily used for feeding the liquid food or food readily soluble in saliva. The following parts may be distinguished:

(i) *Labrum:* It is long tapering and highly sclerotized flap which originates from the distal margin of the rostrum and covers the anterior surface of the haustellum. The epipharynx is attached with its lower surface and jointly termed as labrum epipharynx which helps in forming the food channel with the hypopharynx. The mouth is situated medially at the base of the labrum and hypopharynx.

(ii) *Mandibles:* The, mandibles are altogether absent.

(iii) *Maxillae:* They are represented by only a pair of single segmented maxillary palp.

(iv) *Labium:* It is fully developed and modified into a long proboscis. The proboscis can be distinguished into the following three parts:

(a) Rostrum, is also known as basiproboscis which forms the basal part of the labium. It is broad inverted cone like in structure having mostly the flexible membranous walls. There are two small lateral sclerites which support a pair of single segmented maxillary palp.

(b) Haustellum, is also known as mediproboscis which forms the middle portion of labium. It is cylindrical in shape and is demarcated by a distinct long medio dorsal labial groove which works as the food channel. The labrum epipharynx and hypopharynx also rest in it and covers the upper surface. The salivary channel is situated in the hypopharynx. The posterior wall of the haustellum is distinguished by a prominent plate known as mentum or prementum according to Snodgrass and Imms, respectively. It has also been termed as theca which is formed by the spongy muscles.

(c) Labellum or distiproboscis, the distal portion of the labium and is also known as oral disc. It ends into the two spongy broad lobes termed as labella which are separated by a deep channel opening in the labial gutter. The under surface of each lobe is marked by the numerous (30-32) transverse channels known as pseudotracheae. They have the superficial resemblance to half open tracheal tubes and do not take part in respiration. These pseudotrachea are parallel to each other and open in a collecting canal.

Feeding Mechanism: In the resting stage the proboscis remains retracted under the head, but when the insect is to feed the proboscis is protruded out due to the pressure of blood and head muscles are thrust into the liquid food. This food is received by the pseudotracheae by their capillary action and is taken to food channel from where it finally goes to digestive tract. Certain solid foods such as sugar etc. are also eaten by dissolving them in the saliva and this solution is then drawn up into the food channel. Some time direct feeding is also accomplished by raising both the lobes of labella upward and this facilitates the insect to take the solid (sugar crystals) or liquid food directly into the mouth cavity through prestomum.

Siphoning Type of Mouth Parts

This type of mouth parts is illustrated by the condition met in the butterflies and moths. They are mainly used for sucking nectar from the flowers. With the few exceptions (fruit sucking moths) insects having such type of mouth parts are harmless to crops. These mouth parts consist of the following parts:

(i) *Labrum:* It is not fully developed and represented by a narrow transverse band at the lower edge of the clypeal region of the face. There is a pair of small hairy lobes known as pilfers on the lateral sides of labrum and in its median region epipharynx is demarcated by a small triangular structure which is pointed anteriorly.

(ii) *Mandibles:* In most species they are absent, but the rudiments of mandibles occur in some macro-lepidopterous moths as small immovable lobe.

(iii) *Maxillae:* They are represented by rudimentary palpi and the modified galeae. The galeae are greatly elongated and joined to form a slender hollow tube like proboscis which is coiled up under the head, like a watch spring when not in use. The lacinea is rudimentary or wanting. The proboscis is composed of two lateral pieces which are held together by interlocking grooves and ridges. In this manner the combined grooves form a tube through which liquid food is drawn up to mouth. Besides, innumerable tiny short muscles also extend from one ring of the tube to another inside each half of the proboscis, which probably help in the coiling and uncoiling mechanism of the proboscis. On either side of the proboscis base a small maxillary palp is situated.

(iv) *Labium:* It is undeveloped and represented only by the large, hairy or scaly, three segmented labial palp and a very small basal plate.

(v) *Sucking Pump:* In this type also the sucking pump is well developed and its working is the same as described in the piercing and sucking type of mouth parts.

Feeding Mechanism: The feeding is accomplished by uncoiling the proboscis and projecting its tip into some exposed liquid and then sucking the liquid up through the food channel, which runs full length through the proboscis, thereafter the proboscis is retracted underneath the head. The theories regarding the coiling and uncoiling mechanism of the proboscis are summarised below:

(i) *Fluid—pressure Theory:* According to Schimitt (1938) proboscis is extended by the blood pressure created in the stipes of each maxilla and simultaneously 2 or 3 pairs of cranial muscles (extensor) also help in the coiling process of proboscis.

(ii) *Pradhan and Aeron's Theory (1941):* According to them the oblique muscles help in extending the proboscis, while the uncoiling process is governed by the chitinous and elastic walls of the proboscis.

(iii) *Eastham and Essac Theory (1955):* These scientists were of the opinion that uncoiling of proboscis is governed by the dorsal transverse convexities, which is due to primary intrinsic muscles and haemocoelic turgidity.

(iv) *Srivastava and Pandey Theory (1966):* Contrary to the Eastham and Essac theory, they suggest that coiling of proboscis is because of the intrinsic oblique muscles and haemocoelic fluid is in no one related with this process. They uphold the views of Pradhan and Aeron's theory.

Chewing and Lapping Type Mouth Parts

This type of mouth parts are found in the insects of Order Hymenoptcra, *e.g.*, honey bees and wasps. These are generally used for collecting the pollen, molding the wax or nest material and lapping the nectar from the flowers. The labrum and mandibles are just like chewing type of mouth parts, while the labium and maxillae are modified into proboscis. In such type of mouth parts, the following parts may be distinguished:

(i) *Labrum:* It is a small semicircular lobe found just below the clypeus, having no proper functioning and is almost vestigeal.

(ii) *Mandibles:* On the lateral side of labrum, a pair of mandibles are found, which are more or less flattened or spoon shaped mainly adapted for molding the wax and for comb formation. Besides, they also provide support to the proboscis.

(iii) *Maxillae:* These are paired structures situated on the either side of the labium. In each maxilla the lacinea is either rudimentary or absent, whereas the galea is large, flat, blade like in structure and longer than the stipes. The palpi are vestigeal and small peg like in structure. The cardo is long, peg shaped and between each cardo there is a small 'V' shaped sclerite known as lorum, which laterally articulates with the distal end of the cardo and supports the base of labium. Each maxilla bears a long flattened galea having a vein which joins the subgalea at the base.

(iv) *Labium:* It is distinctly differentiated into mentum, prementum, glossa, paraglossa and labial palp.

The prementum is a long region having a small triangular mentum at its base. The apex of mentum fits with the concavities of the lorum. The glossae are modified into a hairy tongue like structure which is greatly lengthened and highly mobile being flexible in all directions and capable of an active movement of protection and retraction. It terminates into a small lobe known as labellum or flabellum. The lower surface of the tongue is marked by a deep channel, which extends from flabellum to the base of tongue. On the base of either side of tongue, there is a pair lobes known as paraglossa and they cover the channel and opening of the salivary duct, which is present on the dorsal side of the tongue. Thus, paraglossa helps in forcing the saliva to reach the end of hypopharynx where it mixes with the food. The labial palpi are long, four segmented and cover the tongue from the lateral side during the lapping of nectar from the flowers.

Feeding Mechanism: The mechanism of feeding explained by different scientists is given below:

(i) *Snodgrass View:* While feeding, the broad maxillary galeae and the labial palpi are brought together over the tongue thus forming a tubular proboscis and the end of which is thrust into the liquid food by a rapid back and forth movement of tongue, the liquid is drawn up into the tube and where from it is sucked up into the mouth by stomodeal pump. According to Snodgrass glossal canal does not play any part in sucking up the sap.

(ii) *King's Theory:* Contrary to Snodgrass theory George E. King found that first of all tongue (glossa) is thrust into the flower and the nectar sticks with its tip. The tongue, thus smeared with nectar is rapidly retracted between the labial palp and the nectar is squeezed off the tongue by the galeae and deposited in the small cavity formed by the paraglossa and the base of glossa. Then by bending of the labium upward near mid length, the accumulated nectar is sucked into the oesophagous by the action of sucking pump.

(iii) *Imms Theory:* According to him the liquid food is drawn up by the glossal canal through the capillary action and with the muscular action, the glossa is contracted, as a result of that the liquid is sucked up in the mouth through the channel formed by the glossa and paraglossa.

(iv) *Jucker-Piedallus Theory:* When the insects feed on some solid food like sugars, the hard spines located on the labellum of tongue scrap the sugar particles and mix it with the saliva. Thus the liquid food is sucked up in the mouth through glossal canal based on the earlier theory.

Rasping and Sucking Type of Mouth Parts

These are also termed as Thysanopterous mouth parts and are well illustrated by the thrips. Underneath the head a cone shaped proboscis is situated which is formed by clypeus, labium, labrum and the maxillae. It is asymmetrical in shape. In such type of mouth parts the right mandibles is wanting or vestigeal. When a thrip feeds, the apex of mouth cone is applied to the surface of plant and stylets are inserted into the tissues. The laciration of tissues causes a minute wound through which the sap comes out. The apex of the mouth cone is applied to the puncture and the juice of plant is pumped up into the gut by means of the action of cibarial sucking pump.

Degenerate Type: Such type of mouth parts are generally met in the larvae or maggot of Order Diptera and the classical examples are the maggots of fruit flies and housefly. They are poorly developed and represented by only a hypopharyngeal sclerite attached with a pair of hooks which also help in their locomotion.

Thorax

It is the body region situated between head and the abdomen. In certain insect orders like Orthoptera, Dermaptera and Odonota, there is a very small region between the thorax and head known as cervix or collum. Some times it has also been referred as microthorax or the neck region of the insect. On the base of the cervix insect can move its head in different directions. Practically thorax may be termed as the centre of the locomotory organs because the legs and wings, which are considered as the primary organs of locomotion are situated in the thoracic region. Thorax may be differentiated into the following parts:

(i) Prothorax (ii) Mesothorax (iii) Metathorax

Each segment bears a pair of legs. The first pair of wings is attached in the mesothorax, where as the second pair is found in the metathorax. In winged insects the meso- and meta-thorax are so closely attached to form a pterothorax. The thorax is armed with various sclerites, which are termed by their specific locations as tergum or notum situated on the dorsal surface; pleuron located on the lateral sides and the sternum which is invariably present on the ventral surface of thorax.

Prothorax: It is the anterior most region of the thorax which is exceedingly large in grasshoppers and covered by a conspicuous sclerite known as pronotum. It is triangular and covers the mesothorax from its posterior region which can further be differentiated into pre-scutum, scutum, scutellum and post-scutellum. In grasshoppers the sternum is termed as prosternum and its pleural plates are triangular in shape located just above the base of legs.

Meso and Meta-thorax: The meso and meta thorax are almost united to form a compact body so as to accommodate the musculature necessary to combine both running and flying mechanism in one segment. These segments are collectively known as pterothorax. In grasshopper the pleuron consists of an anterior sclerite or episternum and a posterior sclerite or epimeron, which are separated by the pleural suture. The sturnum is composed of a broad flat plate, which is located between the meso and meta-thoracic legs and termed as the mesosternum and metasternum, respectively. There are two pairs of spiracles, the first pair is located between the prothorax and the meta-thorax known as mesothoracic spiracles and the second pair is found above the second pair of legs termed as metathoracic spiracles.

Thoracic Appendages

The thorax consists of the following appendages: (1) Legs (2) Wings

Legs

An insect usually bears three pairs of legs which are located on the ventral surface of thoracic segments. They are the primary organs for running or walking, but according to the habit and habitat of insects they are modified for different purposes. A typical leg consists of the following parts:

(i) Coxa—is the basal segment and joint the leg with thorax. The coxa fits in the cup depression of the body and is generally freely movable.

(ii) Trochanter—is the small second division of the leg which articulates with coxa but is usually fixed with femur.

(iii) Femur—is the largest and most powerful division of the leg.

(iv) Tibia — is the slender, usually quite long and provided with downward projecting spines which helps in climbing.

(v) Tarsus—is the fifth divisional segment and generally sub-divided into 2 to 5 segments, but these segments are devoid of muscles and incapable for movement.

(vi) Pretarsus—is the last terminal segment of the leg which is represented by a complex set of claws which bear one or more pads or cushions at their base or between them. A pad between the claws is known as aerolium, whereas the pads at their base are termed as pulvilli. In most dipterous insects aerolium is situated between the pulvilli which is also some times modified into bristle like strúcture known as empodium.

Modifications of Legs

As discussed earlier the legs of insects are variously modified for the purposes for which they are used. The important modifications of the insect legs are given below:

(i) *Ambulatorial or Walking Type:* It is the generalized form of an insect's leg which is usually adapted for walking. In such type of legs the coxa is smaller, simple and the trochanter is subdivided into two segments. The femur is slender and flat, where as the tibia comparatively longer and adorned by small hard spines on both the sides. The tarsus is four or five segmented whereas the pretarsus is separated by claws and the Pulvilli, e.g., cockroach and bugs.

(ii) *Cursorial or Running Type:* It is almost similar to walking type of leg, but is differentiated by the tarsus which is comparatively longer and touches the ground while running. This adaptation avoids the danger of slipping away while running on the ground, e.g., earwigs, ants and ichneumon wasp.

(iii) *Scansorial or Clinging Type:* Such type of legs are smaller and flat. The coxa are widely separated and one or two segmented tarsi bear a single claw. This claw along with the tarsus is closely attached with the terminal end of the tibia by the muscular action and thus helps the insect to remain attached with the host, e.g., head louse.

(iv) *Saltatorial or Jumping Type:* This type of leg is well illustrated in grasshoppers, crickets and flea beetles where the femur of the hind leg gets enlarged and accommodates the powerful tibial muscles. These muscles help in jumping the insect by their repeated contraction. The tibia is also long and provided with strong and hard spurs at its terminal end. The tarsus is three segmented and they are termed as pretarsus, mesotarsus, and metatarsus, respectively.

(v) *Stridulatorial or Sound Producing:* These legs are typically adapted for producing sound wherein the femur of hind leg of male grasshopper or cricket is provided with the row of pegs (file) on its inner side. These femoral pegs work against the outer surface of each tergum or costal margin of the fore wing thereby producing a sound.

(vi) *Fossorial or Digging Type:* In certain insects the fore legs are modified for digging purpose. They are exceptionally powerful broad and small in size. The coxa, trochanter and femur are strong and broad where as the tibia is provided with four strong scraper like prolongations at its terminal end which are used for digging the ground and cutting the roots of the plants. On the outer surface of tibia, there is ear slit which is meant for hearing. The tarsus is 2 or 3 segmented and each segment projecting like a tooth and helps the tibia for digging the soil, e.g., Mole cricket, Nymphs of Gryllotalpa and Cicada, Grubs of Scarabaeids and Carabids.

(vii) *Natatorial or Swimming Type:* These legs are found in insects living in water and help them in swimming. The femur, tibia and tarsus are flattened and possess the long rows of hairs. The coxa and trochanter are simple and smaller whereas the tarsus is five segmented and the last terminal segment is triangular and pointed. Such type of legs are met in Dytiscus and Jaint water bug.

(viii) *Foragial or Pollen Collecting Type:* This type of modification is found in the legs of worker honey bees which is mainly adapted for carrying the pollen from the flowers. The coxa, trochanter and femur are normal in shape, while the tibia of hind leg is dilated and covered with long dense hairs thus forming a pollen basket or corbicula which is primarily meant for storing the pollen grains. The first tarsal segment (basitarsus) is flattened and its inner surface is densely clothed with several rows of short stiff spines forming a brush known as scopa. It helps the bees in collecting the pollen adhering to the hairs of its body.

(ix) *Raptorial or Grasping Type:* Such legs are adapted for catching the prey and are found in mantis. The coxa of fore leg is more elongated and some times attain the length of femur, whereas the trochanter is small. The femur is thickly spined and groved along their lower side. The tibia is also spinous and fits to the groove along the femur. The prey is thus secured between the grooves and is killed by the spinose femur and tibia. The tarsus is long and five segmented.

(x) *Climbing or Sticking Type:* These are also termed as adhesive type of legs which are generally found in house fly. In such type of legs the pretarsus is highly modified and represented by a pair of claws and a pair of pad like structures known as pulvilli which are found at the base of claws. They are densely clothed with numerous hollow and tubular hairs which secret a sticky substance. Besides, bristle like structure known as empodium is also present in between the pulvilli.

(xi) *Clasping Type:* Such type of legs are modified for reproduction purposes. In these legs the coxa and trochanter are normal and comparatively smaller where as the femur is thick. The tibia is slender and directed towards femur from its inner side. The tarsus is one segmented and arched. During reproduction this adaptation enables the male to hold the female easily, e.g., chleredryinus a parasite of sugarcane pyrilla.

(xii) *Suctorial and Sucking Type:* These legs are also modified for reproduction purposes in which the coxa and trochanter of the fore legs are small, while the femur is thick and small. The tibia is also comparatively smaller and flattened. The tarsus is five segmented wherein the first three segments are modified into a swollen ball like structure and provided with stiff spines at their margins. Besides, there are two vacuum cups on the first tarsal segment which help the male in holding the female during reproduction. They are also helpful in sticking with grasses and leaves etc. against the water-flow, e.g., male Dytiscus.

Wings

The presence of wings is one of the most characteristic feature of the insects and provide one of the most useful parameters for the purpose of classification. Generally, there are two pairs of wings which are the out growth of the body wall along the lateral margins of the thorax. However, in certain insect orders (Thysanura, Protura, Diplura and Collembola) they are altogether absent which is a primitive character of Apterygota (wingless), while in order Diptera there is only one pair of wings and the hind pair is modified into halteres. In some pterygote insects (winged) like ants, termites, bed bugs and the head louse, the loss of wings is a secondary acquired character.

Development of Wings

The wings may be defined as flattened double layered expansion of the body wall which is composed of the same constituents as the body wall namely cuticle, epidermis and basement membrane, while its lumen consists of nerves, tracheae and the blood. The development of wings can easily be seen in the insects having simple or incomplete metamorphosis because they are developed externally. At first the wing buds appear as hollow flattened out growth of body wall along the lateral margin of the dorsum in the meso- and meta-thorax of the second or third instar nymphs. These buds are attached with the body wall which are two layered and contain numerous floating tracheae in the lumen. With the development of wings the wing cavity (lumen) gets gradually shortened and at many places the basement membrane of both the sides gets united, thereby leaving only the wing cavity towards the tracheae.

Thereafter, epidermal cells form the thick cuticular layer along the tracheae for developing the wing veins. Thus, each wing vein is embarrassed with branches of tracheae, nerves and filled blood. When the wing is fully developed the epidermal layer of both the sides also gets united and remains left only around the tracheae. As a result of these changes the upper and lower cuticular layers are also get united and form the cuticular appendages. Since the epidermal cells run along the veins, hence the cross veins also arise from them.

Modification of Wings

The insect wings are highly modified into various forms according to their habits and use hence the distinct names have been applied to those forms:

(i) *Tegmina:* This condition is found in the fore wings of order Orthoptera, Dictyoptera and Phasmida where they are hardened and leathery in consistency.

(ii) *Elytra:* In the insects of order Coleoptera and Dermaptera the four wings get much hardened to form horny sheath which protect the membranous hind wings.

(iii) *Hemelytra:* This is well illustrated in hemipterous insects wherein the fore wings are thickened and their bases like elytra and remaining as soft or membranous that is why they are frequently termed as hemelytra.

(iv) *Halteres:* The hind wings of Diptera are modified into knobbed, thread like balancing organs known as halteres or balancers.

(v) *Pseudohalteres:* These are the reduced fore wings of Strepsiptera.

(vi) *Brachypterous:* In certain grasshoppers the wings are short and this short winged condition is known as brachypterous form.

Wings Margins, Angles and their Regions

In most of the insects the wings are more or less triangular, thereby presenting the three margins; the anterior margin or costa, the outer or apical margin and the inner or anal margin. Besides, three well demarcated angles are also distinguishable, *viz.,* the humeral angle at the base of costa; the apex or angle between the costa and outer margin and the anal angle or tornus between the outer and inner margins.

A typical wing consists of the following regions:

(i) *Axillary Region:* This area is located at the base of the wing. The axillary sclerite is also situated in the same region.

(ii) *Remigium:* This is anterior stiff or hard portion of the wing which is separated by the venal fold from the flexible region of the wing.

(iii) *Vannus:* This region lies between the jugal and the venal fold.

(iv) *Jugum:* This is a lobular region situated near the jugal and the venal fold.

Wing Venation

The complete system of veins of a wing is termed as venation or neuration. Generally, in all the insects there is some similarity in wing venation and therefore, it is presumed that all types of wing venation have developed from the common base or the same ancestor. By means of an extensive study of wing venation in different groups of insects, Comstock and Needham constructed a hypothetical type of veins venation from which all other types have presumely been derived. According to them the primitive wing venation has developed from two tracheae which are situated on the anterior and the posterior basal margins of wings and their branches are spread all over the wing. Each main trachea give rise to three principal veins, thereby forming six principal veins namely costa, radius, medius, cubicus, penultimate and ultimate. Each principal vein gives rise to a sub-vein near its base. The principal veins are represented by + sign whereas the sub veins by - sign. Thus the whole wing venation system is represented by + and the – signs in alternate. The branching of principal veins is represented in the following manner:

Such type of hypothetical wing ventaion is never met in any insect as one or the other vein is invariably found lacking, for example, the medius vein is absent in order Hemiptera and Ephimereda and submedius is missing in Odonata. Some of the scientists consider precosta, costa, subcosta, radius, medius, cubitus and anal as the principal veins of the insect wing.

Cross Veins: The veins joining the two longitudinal veins are known as cross veins. The important cross veins along with their symbols are given below:

(i) *Humeral cross vein (h):* It extends from costa to sub-costa near the humera angle of the vein.

(ii) *Radio-medial cross vein (rm):* It joins the sub-radius and the medius veins.

(iii) *Medial cross vein (m):* The vein joining the m^2 and m^3 branches of medius is termed as medial cross vein.

(iv) *Medio-cubital cross vein (m-cu):* It joins the medius and the cubitus longitudinal veins.

(v) *Radial cross vein (r):* It extends from R_1 to R_2.

Wing Coupling Apparatus

In certain insects special structures have been developed to fasten together the two wings of each side so that it may bring more synchronus action of the fore and hind wings, thereby enabling the insects to fly more swiftly. This action in many insects is ensured simply by fore wing overlapping the hind wing. The important coupling devices developed in insects wing for adding more efficiency in flying are described below:

(i) *Jugal and Humeral Lobe:* This coupling device is commonly found in Lepidoptera, Trichoptera and Mecoptera wherein the wing bases are highly modified. The posterior end of the fore wing is modified into slender finger like organ which is stiffened by a branch of IIIrd anal vein is known as the jugal lobe; whereas the anterior margin of the hind wings is modified in to a small humeral lobe. The lobes of fore and the hind wings are coupled with each other during flight.

(ii) *Frenulum and Retinaculum:* This type of coupling apparatus is well illustrated in higher Lepidoptera wherein the jugum is lost and the frenulum assumes more importance. In female butterflies a number of stout bristle arise beneath the extended fore wing known as frenulum which engages in a retinaculum from a patch of hair near the cubitus of a hind wing. However, in males the frenulum bristles are fused into a single stout structure and is held by a curved process from the sub-costal vein of the fore-wing.

(iii) *Hamuli:* In this modification the costal margin of the hind wings bears a row of small hooks known as hamuli. These hooks get attached into a fold on the inner margin of the fore-wings. Such coupling apparatus is generally met in Hymenoptera.

(iv) *Amplexiform:* This example is commonly met in the insects belonging to family papilionidae and bombycidae of order Lepidoptera. In this case the wings are coupled simply by overlapping basally to each other.

Abdomen

This is the third and the posterior region of the insect body which is comparatively simpler in structure than the thorax. The abdomen is generally comprised of 10 or 11 ring like segments which are also known as uromeres. They are almost similar in structure except of those on the posterior end which are highly modified to form the external genitalia. The abdomen of grasshopper consists of 10 segments, the dorsal surface is composed of a thick cuticular layer known as tergum which forms the 2/3 portions of each of the segment and termed as sternum.

The tergal and sternal plate of each segment is connected with a thin membrane from its lateral regions known as pleura. On each side of the pluron one pair of spiracles is found except on 9th and the 10th abdominal segments. Thus, in all, there are 10 pairs of spiracles; 2 pairs are located in the thoracic regions and remaining 8 pairs in the abdominal region. Besides on each lateral side of the first abdominal segment a tympanal organ is situated which is demarcated by a very thin cuticular layer.

In male and the female grasshoppers 9th and 10th abdominal tergites are so closely united that they appear to be one segment and thus apparently forming 10 abdominal segments. The last abdominal tergite extends posteriorly to form a large plate known as supra-anal plate which has also been considered as the 11th tergite by certain morphologists. The number of sternites is invariably lesser than the tergites; the first eight sternites are distinct, the ninth is membranous while the rest are modified to form the external genitalia in both the sexes. Beneath the supra anal plate, there is a pair of podical plates and a pair of cerci and the anus is situated in between them. According to some scientists the podical plates and anal cerci are the appendages of the 10th sternite.

Usually abdomen serves as a container for the various internal organs of the insects and is the main site of the body responsible for producing the movements of respiration and the process of oviposition.

Abdominal Appendages

The abdominal appendages are non-segmented and may be categorised in to two groups; the first group consists of those appendages which are related with anus and term as anal appendages or the non-reproductive appendages. They are similar in both the sexes and include a large supra anal plate, a pair of podical plates and a pair of anal cerci, while the second group comprises of those appendages which are related with the genitalia openings known as genitalia or the reproductive appendages. Morphologically, they differ in both the sexes and include the appendages of 8th and 9th segments in case of female but only of the 9th segment in case of male.

Female Genitalia

The external genitalia of female grasshopper consists of a copulatory pouch and a special egg laying organ known as ovipositor. These organs are considered to be the appendages of 8th and 9th abdominal segments. The ovipositor helps the female in depositing her eggs in the soil by digging the hole in the ground. There are three pairs of valves which collectively form the ovipositor; the first pair originates from the ninth sternum is triangular, pointed and hard in structure known as dorsal valves, second pair of ventral valves originating from the 8th sternum is similar to dorsal valve and projected upward and a third pair of valves lying in between the dorsal and ventral valves known as inner valves. The inner valves are considered to be vestigial and non-functional whereas the dorsal and the ventral valves fit together to form the functional ovipositor. Besides, an egg guide is also found originating from the base of 8th sternum and extending as a slender appendage through the ventral valves to the genital chamber. Some authors have considered it as a non-functional and vestigal organ.

Male Genitalia

Unlike female genitalia, the genital organs of male grasshopper are located only in 9th segment. The 9th and 10th abdominal tergites are partly fused to form a large cuticular supra-anal plate besides a pair of podical plates and a pair of anal cerci, as seen in the female sex, are also present in the male grasshopper.

The ninth sternum of male grasshopper forms the genital apparatus which is distinctly differentiated in to two parts, first the proximal or the true normal sternite and the second distal, which works as the sub-genital plate and is formed by the union of the coxites of this segment. Only because of this fact the subgenital plate is also known as gonocoxite. The posterior region of the abdomen is thus enclosed directly by supra-anal plate, laterally by podical plates and ventrally by the sub-genital plates. These plates and the aedeagus which are the proximal and distal ends, respectively. All these structures constitute a sperm ejection pump in which opens the ejaculatory duct on the dorsal side of the phallobase, a large complex sclerite is found known as pseudo-sternite which forms an important organ for copulation while in its lateral side two projected parameres are found. In normal condition (at rest), the sub-genital plate presses against the supra anal plate and thus closing the end of the abdomen. In this situation the penis remains folded in the genital atrium.

During copulation the valves of the female genitalia are raised upwards and the male holds the female with the help of parameres and pseudo-sternite. As a result of that the genital armature of both the sexes are engaged with each other. In the male the sub-genital plate is freed and the penis is extended outside. The opening of spermathical duct is now nearly opposite to the penis and receives the spermatic fluid through the vagina which is finally collected into spermatheca.

BIBLIOGRAPHY

Adam, N. : *Encyclopaedia of Plant Physiology,* Springer Verlag, Berlin, 1975.

Alam, B. : *Advanced Plant Physiology,* Longman, London, 1984.

Anderson, Edgar : *Plants, Man and Life,* University of California Press, California, 1967.

Arvill, R. : *Man and Environment : Crisis and the Strategy of Choice,* Penguin, New York, 1967.

Bailey, L. H. : *Hand Book of the Multiplication of Plants,* Discovery Publishing House, New Delhi, 1999.

Beevers, L. : *Nitrogen Metabolism in Plants,* Edward Arnold, London, 1976.

Bhatia, B. M. : *Poverty, Agriculture and Economic Growth,* Vikas Publishing House, New Delhi, 1977.

Billihgs, W. D. : *Plant, Man and the Ecosystem,* MacMillan, London, 1971.

Bowling, D. J. F. : *Uptake of Ions by Plant Roots,* Chapmann and Hall, London, 1976.

Bray, C. : *Nitrogen Metabolism in Plants,* Longman, London, 1984.

Candolle, Alphonse De : *Origin of Cultivated Plants,* Hafner, New York, 1886.

Crafts, A. S. and C. E. Crisp, : *Phloem Transport in Plants,* W.H. Freeman, San Francisco, 1973.

Dadiani, N. K. : *The Survey of Indian Agriculture*, Kasturi & Sons Ltd., Chennai, 1999

Deckock, P. C. : *Mineral Nutrition of Plants : a Symposium*, University of Oxford, New York, 2002.

Dennis, D. T. and Turpin, D. H. : *Plant Physiology, Biochemistry and Molecular Biology*, Longman, London, 1990.

Douglas, J. S. : *Advanced Guide to Hydroponics*, Pelham Books, London, 1976.

Dubey, R. S. : *Agricultural : Issues and Application*, Gyan Books, New Delhi, 1987.

Dunn, Edgar S. : *Location of Agricultural Production*, University of Florida Press, New York, 1954.

Evans, L. T. : *Day Length and the Flowering of Plants*, W.A. Benjamin, California, 1975.

Farmer, J. B. : *Study of Plant Life*, Discovery Publishing House, New Delhi, 1993.

Frankel, F. R. : *India's Green Revolution : Agriculture Costs of Land Growth*, Princeton University Press, Princeton, 1971.

Ftzler M. E. : *Plant Lectins : Molecular and Biological Aspects*, Academic Press, New York, 1985.

Gangawane, L. V. : *Vistas in Mycology and Plant Pathology*, Commonwealth Publishers, New Delhi, 2000.

Gauch, H. G. : *Inorganic Plant Nutrition*, Hutchinson and Ross, London, 1972.

Green, J. R. : *An Introduction to Vegetable Physiology*, MacMillan, London, 1911.

Gregor, Howard F. : *Geography of Agriculture : Themes in Research*, Prentice Hall, New Jersey, 1970.

Gupta, G.P. : *Test Book of Plant Diseases*, Discovery Publishing House, New Delhi, 2001.

Hansen, E. : *Post Harvest Physiology of Fruits*, Praeger Publishers, New York, 1966.

Hewitt, E. J. and Cutting, C. V. : *Nitrogen Assimilation of Plants,* Academic Press, London, 1982.

Hussain, M. : *Systematic Agricultural Geography,* Rawat Publications, Jaipur, 1999.

Isser, G. P. : *Changing Face of Poverty and Globalisation,* Gyan Books, New Delhi, 2002.

James, A.: *Weather and Agriculture,* Pergamon, Oxford, 1967.

Joshi, H. L. : *Agricultural of India,* Rawat Publications, Jaipur, 1990.

Karanth, Apoorva : *Plant Genetics and Food Crop Science,* Dominant Books, New Delhi, 1999.

Kirkby, E. A. : *Nitrogen Nutrition of Plants,* University of Leeds, Leeds, 1970.

Kramer, A. L. : *Principles of Biochemistry,* Worth Publishers, New York, 1982

Kramer, P. J. : *Plant and Soil Water Relationship : A Modern Synthesis,* McGraw Hill Book Company, New York, 1969.

Kumar, Arun : *Agricultural Development : Today and Tomorrow,* Anmol Publications, New Delhi, 1995.

Kumar, T. : *History of Rice in India : Mythology, Culture and Agricultural,* Gyan Books, New Delhi, 1988.

Laxmi Narasaiah, M. : *Agricultural Production,* Discovery Publishing House, New Delhi, 1999.

Mazliak, P. : *Lipid Metabolism in Plants,* Chapmann and Hall, London, 1973.

McKee, H. S. : *Nitrogen Metabolism in Plants,* Oxford University Press, London, 1962.

McNeil, M. Darvill : *Structure and Function of the Primary Cell Walls of Plants,* McGraw, London, 1984.

McRae, S. C. and Burnham, C. P. : *Land Evaluation,* Oxford University Press, New York, 1981.

Meidner, H. and Sheriff, D. W. : *Water and Plants,* Blackie and Sons, Glasgow, 1976.

Mengel, K. and Kirkby, E.A. : *Principles of Plant Nutrition,* Praeger Publishers, New York, 1974.

Mishra, S. R. : *Bacterial Plant Diseases,* Discovery Publishing House, New Delhi, 2003.

Misra, B., Kar, G. C. and Misra, S. N. : *Agro-Industries and Economic Development,* Deep & Deep Publications, New Delhi, 2002.

Mitter, V. : *Growth of Urban Informal Sector,* Deep & Deep Publications, New Delhi, 1988.

Mukherji, Shekhar : *Land-Use Change, Environment and Sustainable Development,* Banaras Hindu University, Varanasi, 2000.

Nomura, M. : *The Control of Ribosome Synthesis,* Edward Arnold, London, 1986.

Novikoff, A. B. : *The Endoplasmic Reticulum : a Cytochemist's View,* Oxford, New York, 1976.

Numa, S. : *Fatty Acid Metabolism and its Regulation,* Janis, London, 1984.

Nutman, P. S. : *Symbiotic Nitrogen Fixation in Plants,* Cambridge University Press, Cambridge, 1976.

Ogawa, Y. and R. W. King : *Plant and Cell Physiology,* Springer-Verlag, Berlin, 1990.

Opik, H. : *The Respiration of Higher Plants,* Edward Arnold Publishers, London, 1980.

Palladin, V. I. : *Plant Physiology,* Arihant Publishers, Jaipur, 1988.

Pannikkar, K. M. : *India and the Indian Ocean : An Essay on the Influence of Sea Power on Indian History,* George Allen and Unwin, London, 1951.

Ramshaw, J. A. M. : *Structure of Plant Proteins,* Oxford University Press, New York, 1982.

Robert, W. : *Middle America : Its Land and Peoples*, Prentice Hall, Englewood Cliff, 1989.

Roberts, J. A. and Tucker, G. A. : *Ethylene and Plant Development*, Butterworth, London, 2001.

Rubenstein, J. M., and Bacon, R. S. : *The Cultural Landscape : An Introduction to Human Geography*, Prentice Hall, New Delhi, 1990.

Salisbury, F. B. and C. W. Ross, : *Plant Physiology*, Woodworth Publishing Co., California, 1967.

Samanta, R. K. : *Agricultural Estension in Changing World Perspective*, Uppal Publishing House, New Delhi, 1995.

Shandhya, T. K. : *Agricultural Credit and Nabard*, Deep & Deep Publications, New Delhi, 2003.

Sharma, D. N. : *Minerals of India*, National Book Trust, New Delhi, 1994.

Sharma, S. S. : *Sustainable Agriculture : Poverty and Food Security*, Rawat Publications, Jaipur, 2002.

Sharma, V. K. : *Trees and Protection of Environment*, Deep & Deep Publications, New Delhi, 2004.

Siegelman, H. W. and G. Hind : *Photosynthetic Carbon Assimilation*, Plenum, New York, 1978.

Singh, D. K. : *Technological Change and Agricultural Development*, Deep & Deep Publications, New Delhi, 1993.

Singh, Harjinder : *Agricultural Problems in India*, Gyan Books, New Delhi, 1988.

Singh, L. : *Green Revolution and Cropping Pattern*, Deep & Deep Publications, New Delhi, 1993.

Singh, Surendra : *Agricultural Development in India : A Regional Analysis*, Kaushal Publications, Shillong, 1994.

Slatyer, R. O. : *Plant Water Relationship*, Academic Press, New York, 1967.

Somani, K. K. : *Indian Economics and Social Traditions*, Gyan Books, New Delhi, 2002.

Srivastava O. S. : *Agricultural Economics*, Rawat Publications, Jaipur, New Delhi, 1996.

Steward, F. C. : *Plant Physiology, A Treatise*, Academic Press, New York, 1963.

Street, H. E. and H. Opik : *The Physiology of Flowering Plants*, Edward Arnold, London, 1970.

Subrahmanya, S. : *Studies in Agricultural Development*, Deep & Deep Publications, New Delhi, 1996.

Sudhir, M. : *Applied Biotechnology and Plant Genetics*, Dominant Books, New Delhi, 2000.

Thomas, C. : *A History of the Land and Agriculture*, Prentice Hall, Englewood Cliff, 1978.

Valcarcel, M. : *Weather and Agriculture*, Oxford University Press, London, 1997.

Varner, J. E., Flint D. and R. Mitra, : *Characterization of Protein Metabolism in Cereal Grains*, Academic Press, New York, 1976.

Venkateshwarlu : *Developing Agricultural Technology*, Rawat Publications, Jaipur, 1998.

Verma, M. : *The Survey of Indian Agriculture*, Kasturi & Sons, New Delhi, 1999.

Verma, P. C. : *Surplus Manpower in Agriculture and Employment Policy*, Deep & Deep Publications, New Delhi, 1991.

Vyas, V. S. : *Policies for Agricultural Development*, Rawat Publications, Jaipur, 1997.

Wilkins, M. B. : *The Physiology of Plant Growth and Development*, McGraw Hill book, London, 1969.

Zeevart, J. A. D. : *Physiology of Flower Formation*, McGraw Hill, London, 1976.

Zimmermann, M. H. : *Encyclopeadia of Plant Physiology*, Academic Press, New Work, 1982.

P

INDEX

A

B

C

D

E

F

G

H

N

O

Index

C

D

E

F

G

H

I

J

L

M

N

O

P

Q

R

S

T

U

V

W

Z

❑❑❑